Protein Detection

Edited by Yusuf Tutar and Lütfi Tutar

Published in London, United Kingdom

Protein Detection
http://dx.doi.org/10.5772/intechopen.95147
Edited by Yusuf Tutar and Lütfi Tutar

Contributors
Timothy Veenstra, Joshua Yu, Shenbagamoorthy Sundarraj, Gopalan Rajagopal, Madasamy Sundar, Balaji Sundaramahalingam, Ramar Thangam, Ohanube A.K. Goodluck, Obeta Mark M. Uchejeso, Ikeagwulonu R. Chinaza, Yusuf Tutar, Elif Cansu Abay, Mervenur Al, Nazlıcan Yurekli, Merve Tutar, Kubra Acikalin Coskun, Nilam bhasker

First published in London, United Kingdom, 2022 by IntechOpen
IntechOpen is the global imprint of INTECHOPEN LIMITED, registered in England and Wales, registration number: 11086078, 5 Princes Gate Court, London, SW7 2QJ, United Kingdom
Printed in Croatia

British Library Cataloguing-in-Publication Data
A catalogue record for this book is available from the British Library

Additional hard and PDF copies can be obtained from orders@intechopen.com

Protein Detection
Edited by Yusuf Tutar and Lütfi Tutar
p. cm.

This title is part of the Biochemistry Book Series, Volume 31
Topic: Proteomics
Series Editor: Miroslav Blumenberg
Topic Editors: Paolo Iadarola and Simona Viglio

Print ISBN 978-1-83969-352-6
Online ISBN 978-1-83969-353-3
eBook (PDF) ISBN 978-1-83969-354-0
ISSN 2632-0983

For EU product safety concerns:
IN TECH d.o.o., Prolaz Marije Krucifikse Kozulić 3, 51000 Rijeka, Croatia,
info@intechopen.com or visit our website at intechopen.com.

IntechOpen Book Series
Biochemistry
Volume 31

Aims and Scope of the Series

Biochemistry, the study of chemical transformations occurring within living organisms, impacts all of the life sciences, from molecular crystallography and genetics, to ecology, medicine and population biology. Biochemistry studies macromolecules - proteins, nucleic acids, carbohydrates and lipids –their building blocks, structures, functions and interactions. Much of biochemistry is devoted to enzymes, proteins that catalyze chemical reactions, enzyme structures, mechanisms of action and their roles within cells. Biochemistry also studies small signaling molecules, coenzymes, inhibitors, vitamins and hormones, which play roles in the life process. Biochemical experimentation, besides coopting the methods of classical chemistry, e.g., chromatography, adopted new techniques, e.g., X-ray diffraction, electron microscopy, NMR, radioisotopes, and developed sophisticated microbial genetic tools, e.g., auxotroph mutants and their revertants, fermentation, etc. More recently, biochemistry embraced the 'big data' omics systems. Initial biochemical studies have been exclusively analytic: dissecting, purifying and examining individual components of a biological system; in exemplary words of Efraim Racker, (1913 –1991) "Don't waste clean thinking on dirty enzymes." Today, however, biochemistry is becoming more agglomerative and comprehensive, setting out to integrate and describe fully a particular biological system. The 'big data' metabolomics can define the complement of small molecules, e.g., in a soil or biofilm sample; proteomics can distinguish all the proteins comprising e.g., serum; metagenomics can identify all the genes in a complex environment e.g., the bovine rumen.
This Biochemistry Series will address both the current research on biomolecules, and the emerging trends with great promise.

Meet the Series Editor

Miroslav Blumenberg, Ph.D., was born in Subotica and received his BSc in Belgrade, Yugoslavia. He completed his Ph.D. at MIT in Organic Chemistry; he followed up his Ph.D. with two postdoctoral study periods at Stanford University. Since 1983, he has been a faculty member of the RO Perelman Department of Dermatology, NYU School of Medicine, where he is codirector of a training grant in cutaneous biology. Dr. Blumenberg's research is focused on the epidermis, expression of keratin genes, transcription profiling, keratinocyte differentiation, inflammatory diseases and cancers, and most recently the effects of the microbiome on the skin. He has published more than 100 peer-reviewed research articles and graduated numerous Ph.D. and postdoctoral students.

Meet the Volume Editor

Prof. Dr. Yusuf Tutar conducts his research at the Hamidiye Faculty of Pharmacy, Department of Basic Pharmaceutical Sciences, Division of Biochemistry, University of Health Sciences, Turkey. He is also a faculty member in the Molecular Oncology Program. He obtained his MSc and Ph.D. at Oregon State University and Texas Tech University, respectively. He pursued his postdoctoral studies at Rutgers University Medical School and the National Institutes of Health (NIH/NIDDK), USA. His research focuses on biochemistry, biophysics, genetics, molecular biology, and molecular medicine with specialization in the fields of drug design, protein structure-function, protein folding, prions, microRNA, pseudogenes, molecular cancer, epigenetics, metabolites, proteomics, genomics, protein expression, and characterization by spectroscopic and calorimetric methods.

Dr. Lütfi Tutar is currently an assistant professor at the Department of Molecular Biology and Genetics, Faculty of Art and Sciences, Kırşehir Ahi Evran University, Turkey. His interdisciplinary research focuses on bioinformatics analysis of high-throughput data, microRNAs, small RNAs, and Heat Shock Proteins (HSPs) in human diseases and other multicellular organisms.

Contents

Preface

This book explores distinct aspects of protein purification and characterization steps.

Chapter 1, "Structure- and Design-Based Difficulties in Recombinant Protein Purification in Bacterial Expression", focuses on issues that are not addressed in straightforward purification processes, such as if the peptide tends to form aggregates and displays solubility problems. It also discusses strategies for small peptide chromatography. The chapter provides solutions for solubility and aggregation problems, such as a unique column and resin selection trick. The proper selection of purification parameters helps the purification process.

Chapter 2, "Protein Detection in Clinical Diagnosis and Management of Prevalent Neurodegenerative Diseases and Metabolic Disorders", examines protein-related diseases and the diagnosis of proteinopathy-related diseases considering the structure and function of the proteins through their deficiencies and overexpression. It also discusses the pathogenesis and the principles underlying the diagnosis of these disorders.

Chapter 3, "Methods of Protein Detection in Cancer for Diagnosis, Prognosis and Therapy", highlights enzyme-linked immunosorbent assay (ELISA), immunohistochemistry, flow cytometry, western blot, mass spectrometry, protein microarray, and microfluidics techniques for screening, protein profiling, identification, and qualitative and quantitative analysis of differential expressed oncoproteins at different stages of cancer.

Chapter 4, "Multiple and Single Reaction Monitoring Mass Spectrometry for Absolute Quantitation of Proteins", further elaborates on omic technology. It discusses the technology behind mass spectroscopy-based clinical protein assays in diagnostic and prognostic settings.

Finally, Chapter 5, "Interrelation between Recurrent Pregnancy Loss and Antiphospholipid: A Clinico-Diagnostic Perspective", discusses the role of antiphospholipid in recurrent pregnancy loss from a clinical-diagnostic perspective. Although the most common cause of recurrent pregnancy loss is a cytogenetic anomaly, antiphospholipid antibody, and metabolic and hormonal disorders, half of the cases of recurrent pregnancy loss remain unexplored.

Yusuf Tutar
Hamidiye Faculty of Pharmacy,
Department of Basic Pharmaceutical Sciences,
Division of Biochemistry,
University of Health Sciences-Turkey,
Istanbul, Turkey

Lütfi Tutar
Kırşehir Ahi Evran University,
Kırşehir, Turkey

Section 1

Strategies and Methods for Protein Detection

Chapter 1

Structure- and Design-Based Difficulties in Recombinant Protein Purification in Bacterial Expression

Kubra Acikalin Coskun, Nazlıcan Yurekli, Elif Cansu Abay, Merve Tutar, Mervenur Al and Yusuf Tutar

Abstract

Protein purification is not a simple task. Yet, overexpression at bacterial systems with recombinant modifications brings further difficulties. Adding a tag, an affinity label, and expressing particular domains of the whole protein, especially hydrophobic sections, make purification a challenging process. Protein folding pattern may perturb N- or C-terminal tag and this terminal preference may lead to poor purification yield. Codon optimization, solvent content and type, ionic conditions, resin types, and self-cleavage of recombinant proteins bring further difficulties to protein expression and purification steps. The chapter overviews problems of protein purification through a small peptide overexpression in bacteria (Recombinant anti-SARS Coronavirus 2 (SARS-Cov-2) Spike protein Receptor Binding Domain (RBD) antibody (Clone Sb#14). The chapter also covers troubleshooting at distinct steps and highlights essential points to solve crucial issues of protein purification.

Keywords: protein, prokaryotic protein expression, and purification, protein modeling, protein aggregation, ionic strength

1. Introduction

Recombinant DNA technology involves genetic engineering; cutting DNA molecules from distinct biological species and then ligating them to a vector for expression [1, 2]. The technology helps to express the desired protein in large quantity rather than extracting from bulk amounts of tissues and animal fluids. Proteins are synthesized and modified depending on their functions in an organism. As an initial step, DNA encodes protein through transcription from mRNA synthesis. Then, mRNA is converted into protein. Transcription and translation occur simultaneously in prokaryotic organisms. The conversion of mRNA to protein begins before the synthesis of the mature mRNA transcript [3]. Protein expression involves the synthesis, modification, and regulation of a particular protein in a living organism. However, bacterial systems lack human protein modifications but overexpress recombinant proteins in bulk amounts [4]. Recombinant protein expression is useful to understand the structure and function of proteins. A network of protein complex functions can be distinguished by the characterization of individual proteins function as well as interactions through recombinant protein techniques. Protein–protein and protein-ligand interactions may be highlighted by expressing interacting domains and by introducing key mutations to reveal key domains and residues, respectively. Considering the size and complexity of proteins, protein production is very efficient with vector templates using K12 bacterial systems [2, 4].

1.1 Bacterial protein synthesis

Recombinant protein production in bacterial systems is fast, easy, and highly efficient [5]. The general strategy for recombinant protein production involves transforming the cell with a DNA vector containing the open reading frame of the gene, then after subcloning to the expression system, the protein is induced in the cells. After incubating the induced cells, harvested cells are lysed for further separation. The selection of the purification system depends on the type of protein, the affinity tag of the plasmid, the isoelectric point (pI) of the protein of interest, the molecular mass of the protein, the targeted yield, and the degree of functional activity. The lysed cells are purified through column chromatography with a proper resin(s) and a convenient buffer system [6, 7]. But in practice, several steps may cause problems. These include inadequate growth of the selected bacterial host cell, the formation of inclusion bodies, protein aggregation, structural alteration, recombinant protein nonspecific interaction with cellular proteins, problems in colon systems used in purification [6, 8]. Further, as eukaryotic proteins expressed in E. coli may not perform post-translation modifications of other organism proteins, loss of function is observed. In addition, some of the proteins expressed are exposed to hard-denaturing agents or may cause collapses in structure and this often results in insolubility problems. Overexpression of recombinant proteins in bacterial systems leads to the formation of inclusion bodies. Re-folding of these proteins into their bioactive forms is cumbersome and requires a variety of agents and processes [1, 9].

1.1.1 Host cell selection

First of all, the choice of the host cells to produce intact protein in the synthesis mechanism forms the mainline of the whole system. Microorganisms used in recombinant protein expression systems include bacteria and yeast. Each host has strengths and weaknesses. The organism to be selected varies depending on the particular protein, working conditions, desired yield. For example, if the desired protein has post-translational modifications, choosing a prokaryotic expression system would not be proper [10]. BL21 (DE3) and its derivatives are by far the most commonly used strains for recombinant protein synthesis. In addition, its genetics are characterized in more detail than any other microorganisms. Recent studies suggest that BL21 (DE3) gene-level research made this bacterium more important for the production of heterologous proteins. This host cell provides maximum efficiency in protein expression through inexpensive substrates, capable of rapid and high-yield growth. A modified form includes a pLysS plasmid that encodes T7 lysozyme. This lowers the background protein expression of recombinant protein but does not perturb IPTG induction. The plasmid is especially useful in toxic cases and provides an option for protein over-expression. Yeast is an alternative recombinant protein production host and provides eukaryotic post-translational modifications with high yield.

Yeast growth temperature (30°C) is lower than that of bacteria (37°C) but the growth rate is much slower. Further, the transformation of plasmids to yeast is relatively difficult and the selection of transformed cells and growth conditions require special conditions [10, 11].

1.1.2 Plasmid selection

The expression plasmids consist of the replication origin, promoter, and multiple cloning sites. The most important issue to consider when choosing an appropriate vector is the copy number property. Because the number of copies is controlled by the replication. It is not always true to assume that the high amount of plasmid is proportional to the yield of recombinant protein expression. Because the high copy is inversely proportional to the rate of bacterial growth. In addition, this condition creates plasmid instability and creates a metabolic load. As a result protein production yield decreases [12, 13].

1.1.3 Promoter

Prokaryotes have to adapt to the environment by responding quickly to environmental changes. E. coli cells cannot use lactose directly as a source of carbon. But they use glucose, a component of lactose. For the bacterial cell to metabolize lactose, it is necessary to take lactose into the cell and break it down into a glucose monomer. For this, it is necessary to synthesize three different enzymes in the cell [6, 14]. As with E. coli, bacteria combine genes related to the same metabolic pathways to form clusters called operons. Transcription of the genes that make up the operon start from a single promoter. The resulting transcription product consists of an mRNA molecule containing information from multiple genes. Preserved DNA sequences in the promoter region help connect the enzyme to the DNA molecule. Induction is difficult in the presence of easily metabolized carbon sources. If lactose and glucose are present in the environment, expression from the lac promoter is not fully induced until all glucose is used up. In the absence of glucose, the promoter expresses the three enzymes to break down the lactose to obtain glucose. This property is used to induce prokaryotic expression vectors through IPTG (isopropyl 1-thio-β-d-galactopyranoside); a lactose analog that binds lac repressor [14, 15]. In the commercial vectors, IPTG starts the transcription of the lac operon and eventually induces protein expression where the gene of interest is controlled by the *lac* operator.

1.1.4 Marker selection

A resistance marker is added to the plasmid to prevent the growth of cells that do not carry plasmids. This can be achieved by using a selection marker. For example, ampicillin resistance is conferred by the *bla* gene, β-lactamase, a periplasmic enzyme that inactivates the β-lactam ring of β-lactam antibiotics [16].

1.1.5 Affinity tags and its contribution to protein solubility

The addition of affinity tags to the plasmid (such as His Tag, glutathione-S-transferase, and cellulose-binding domain) is employed to separate a particular protein from the heterogeneous protein mixture during purification, forming disulfide bonds, increasing the solubility of the recombinant proteins and transferring them to the periplasm region. Affinity tags have a great role in separating the desired protein from cell lysate in recombinant protein purification. Affinity tags are divided into small peptide tags (amino acids) and large polypeptide tags (fusion partners) [17]. Small peptide tags are less likely to interfere when fused to the protein. In some cases, this may have negative consequences on the tertiary conformation and biological activity of the fused chimeric protein. Vectors are available that allow tags to be placed optionally at the N-terminal or C-terminal end. It is more advantageous to position a signal peptide at the N-terminal end for better secretion of the recombinant protein. At this point, it is important to know which end of the protein is embedded in the folding pattern by examining the three-dimensional structure of a particular protein, and it is necessary to place the label on the solvent-exposed end. Examples of small peptide tags are poly-His, c-Myc, and FLAG [18]. His-tagged proteins can be purified by affinity chromatography in resins containing positively charged metal ion nickel. In addition, at the end of purification, with commercial antibodies, labeled recombinant protein can be detected by western blot [17–21]. On the other hand, it increases the solubility of the recombinant protein produced by the addition of a non-peptide fusion partner (large polypeptide label). The most commonly used fusion labels include Thioredoxin (Trx), Ubiquitin, SUMO, Maltose binding protein (MBP), Glutathione S-transferase (GST) [17, 22]. The reason why fusion partners show properties that increase the solubility of the protein is still not fully explained. Though, MBP label has been shown to carry a small chaperone activity. The GST label has been shown to have the weakest solubility-enhancing effect among fusion partners. Trx has the most solubility-enhancing properties, but due to its size, it may cause adverse effects. In recent years, studies have shown that "Calcium-Binding Protein Fh8" tag derived from a parasite called "Fasciola hepatica" recombinantly added to proteins increases protein solubility [6, 17, 20, 21]. Studies are underway for better solubility enhancing effect of recombinant protein tags.

1.2 Troubleshooting strategies for recombinant protein expression

Even if the effective parameters are provided in the production of recombinant protein, it may not be determined exactly whether the desired protein will be eluted excessively and in active soluble form. Therefore, there are additional strategies for optimizing protein expression [7].

1.2.1 Low or no protein production

If the desired protein cannot be detected using sensitive techniques or is detected at a low expression rate, the problem is usually caused by a toxic effect of the heterologous protein in the cell. As a result of protein toxicity in the host cell, cells cannot proliferate at a sufficient level and show a low growth rate [7, 23]. The first measure to solve this problem should be followed before proliferating cells are induced. If the growth rate of the recombinant cell is slower than that of the strain with empty vectors, it is related to either gene toxicity or the basal expression of toxic mRNA and protein. Control of basal production is associated with the operon system. *Lac*I or *Lac*IQ expression blocks transcription in *Lac*-based promoters. High-copy plasmids must be cloned in the *Lac*I Q expression vector. Since the presence of tryptone or peptone in the growth medium contains inducing lactose, a more controlled expression is provided with the addition of glucose at 0.2–1 w/v. Plasmids containing T7-based promoters prevent leaky production, such as BL21DE3-pLYS (S) [8, 24].

1.2.2 Limiting factors in the medium

Luria Bertani (LB), the most commonly used growth medium environment for E. coli culture, is an ideal environment for high-nutrient cell growth. When recombinant protein production cannot be replicated with the recommended mechanisms, production efficiency can be increased by increasing the volume of the targeted protein. A successful result can be achieved with adequate ventilation with rigorous shaking of the growth medium. Although LB has a high protein content, cell proliferation is partially reduced. This is due to the low carbohydrate content of LB. As a solution to this situation, increasing peptone and yeast extract provides higher cell proliferation with the addition of $MgSO_4$, which contributes to the sonic intensity of the environment. In addition, the amount of acid released as a result of increased glucose metabolism over time exceeds the buffering capacity of LB. In case of acidification of the growth medium, 50 mM phosphate salts can be added to the environment and buffered [7, 11]. In the broth culture, as the number of cells per unit media increases, oxygen limitation occurs and changes the metabolic capacity of the cell. This prevents optimal growth and the easiest way to increase the amount of oxygen in the growth medium is to increase the speed of the shaking containers. The optimum shaking speed range is 300–400 rpm. Several anti-foaming agents can be added to the broth culture to prevent the negative effect of the foams formed by strong shaking on oxygen circulation [24].

1.2.3 Formation of inclusion bodies

The inclusion bodies formed in E. coli are denatured protein molecules that do not display biological activity. Dissolving, refolding, and purification protocols should be applied, respectively, to make inclusion objects functionally active and soluble. In the transfer of a foreign gene to E. coli, control of gene expression is lost. The nascent polypeptide expression depends on several factors such as osmosis, folding pattern, and pH. If expression increases, the number of unspecific hydrophobic interactions in the polypeptide chain increases. This causes instability and clustering in poly peptization. The resulting protein aggregation is called "inclusion bodies." The main reason for the formation of clustering is due to the deterioration of the balance between protein aggregation and protein resolution [1, 25]. Therefore, a soluble recombinant protein can be obtained through strategies that eliminate the factors causing the formation of inclusion bodies. As mentioned in the "Affinity tags" section, one way to prevent the solubility problem that may occur in the expressed protein is; combining the desired protein with a fusion partner (large polypeptide tag) that acts as a solubilizer [17].

1.2.4 Disulfide bond formation

To obtain the biologically active three-dimensional structure of recombinant proteins, it is important to establish the right disulfide bonds. The formation of improper disulfide bonds causes the protein to fold incorrectly and the formation of inclusion bodies. Disulfide change reactions catalyzed by many enzymes in the Dsb family, where cysteine oxidation occurs in E. coli periplasm, form disulfide bonds in the polypeptide chain [26]. In the cytoplasm, the formation of disulfide bonds is rare because the remnants of cysteine are catalytic regions for many enzymes in the cytoplasm. The wrong disulfide bonds in these regions can cause protein inactivation, clustering, and incorrect folding. However, some strains of E. coli have conditions that trigger the formation of a disulfide bond [5].

1.2.5 Addition of chemical chaperones and co-factors

Molecular chaperones form the heart of protein synthesis and help nascent polypeptides fold into their active structures. Some specific types of chaperones, such as ClpB, can cleave unfolded polypeptides contained in inclusion bodies. However, high levels of recombinant protein production may result in increased molecular traffic in the cytoplasm, resulting in uncontrolled protein folding control. One strategy used to solve this problem is to arrest protein expression by removing the inducer after a centrifugation step and adding a fresh medium containing chloramphenicol, the protein expression inhibitor. Thus, it allows the recruitment of molecular chaperones to enable the folding of newly synthesized recombinant polypeptides [27, 28]. One of the systems used commercially for protein folding is chaperone plasmids. This system consists of plasmids that allow overexpression of different chaperones or their combinations. Examples of these are GroES-GroEL, DNAK/DNAJ/GrpE [27]. When proteins are released from inclusion bodies, denatured with urea, and subsequently folded *in vitro*, the addition of osmolytes (proline, trehalose) at a concentration ratio of 0.1–1M increases the yield of soluble protein. In addition, the correctly folded protein may require special cofactors such as metal ions (such as magnesium, iron/sulfur) or polypeptide cofactors in the media medium to reach its final conformation. The addition of these compounds to the culture increases the yield and the folding rate of soluble proteins [8, 28].

1.2.6 Slowing down the production rate

Slowing the production rate of the recombinant protein reduces cellular protein concentration and protein trafficking, allowing the synthesized polypeptides to fold more smoothly. The most common method of reducing the rate of protein synthesis is to lower the incubation temperature [29]. Decreased temperature prevents the formation of aggregation due to its reduction of hydrophobic interactions. Recombinant protein synthesis occurs in the temperature range of 15–25°C. However, when working at the lower temperature range, this causes slower growth and therefore lower protein synthesis. This obstacle can be overcome with commercial products. The ArticExpress™ (Agilent Technologies) competent cells improve recombinant protein expression at low temperatures through co-expressing ortholog genes of E. coli GroEL and GroES from Oleispira Antarctica, namely Cpn60 and co-chaperone Cpn10. These chaperones work together to fold a substrate protein, and usually carry re-folding activity at 4–12°C temperature range, increasing recombinant protein yield and solubility at lower temperatures [30].

2. Techniques used in recombinant protein purification and detection

Selection of the purification methods generally uses distinct characteristics of the proteins. The distinct properties of recombinant proteins may include chemical, biological, and physical features due to differences in spatial structure and amino acid sequences. Usually, to benefit from these differences, multiple steps are required in the optimal purification process but it should be noted that each step may cause loss of product stability and/or yield, therefore the lowest number of steps are recommended overall for maximum yield. So, method selection determines the ratio of better yield to the better-purified product. Key factors that can affect the purification selection steps include the solubility of the lysate, sample size, and physicochemical properties of the target protein. The first step for purification is to analyze the protein characteristics and match them with literature reports-protocols. For example, a useful parameter in the purification process is amino acid composition. pKa and pI values can be calculated using the amino acid composition. Determination of the values helps to select column type, buffer, pH, or resin type. Once optimization of the purification is established, the method may be employed for a protein with similar sequences or motifs at least for orthologs or isoforms [31]. The characteristic features of proteins that are mainly used for purification type selection are solubility, size, charge, and specific binding affinity [32]. By using these properties, numerous techniques may be employed in protein purification. Solubility parameter can be used with "salting out" through the knowledge of proteins mostly being less soluble in high salt concentrations. And hence, this strategy can be used to separate the protein of interest. Further, dialysis can be used after salting out to remove the salt molecules [33]. Another technique that uses size difference is gel-filtration or size exclusion chromatography (SEC). A column with porous beads resin is used for this and as the sample goes through the column, beads help to separate molecules. The beads are usually 0.1 mm in diameter, so bigger molecules cannot permeate the pores but small molecules penetrate into the pores and are trapped there for a while until the molecules exit again and return to solvent. This action retards small molecules but bigger molecules travel rapidly through a void volume with buffer flow. Small molecules shielding and bigger molecules faster flow separate molecules from each other in fractions depending on their sizes. And as the molecules exit the column, bigger molecules elute first and then, smaller molecules come after.

If the net charge is criteria to be used as a separating feature, ion-exchange chromatography can be used. If the target protein is positively charged as a cationic protein, then, a negatively charged carboxymethyl-cellulose (CM-cellulose) pre-packed column/resin can be used. But if the protein is negatively charged as in anionic proteins, then positively charged diethylaminoethyl-cellulose (DEAE-cellulose) pre-packed columns/resin can be used [33]. It is also known that proteins can have high affinities for certain chemical groups. Affinity chromatography can use this feature to purify proteins and its effect is the best on proteins with affinities to highly specific molecules.

Distinct separation techniques, that is, ion exchange and gel filtration may be employed at high-pressure liquid chromatography (HPLC) with proper column selection. This technique differs from the others because the applied pressure is significantly higher and it does not rely on gravity for sample flow. However, high-pressure limits the purification of higher molecular weight proteins as the pressure denatures protein structure. For higher-molecular-weight proteins FPLC (fast-pressure liquid chromatography) is preferred to prevent pressure-dependent denaturation. FPLC is the preferred technique for protein chemists since any target protein can be separated from cellular lysate readily. And the technique provides a wide range of column options. The flexibility of this technique provides the purification of stable proteins with a high yield. Lower pressure provides advantages as well. Clogging due to lysate content and backpressure problems are less likely encountered compared to that of HPLC. The techniques may also be used with tagged proteins. Histidine tag is one of the most common ones that are used with recombinant proteins and it has a high affinity of metal ions like Ni^{2+} [17]. To screen if the purification steps are working, gel electrophoresis can be used. In-gel electrophoresis, proteins are separated by their mass as they go through the gel and the smaller ones move faster. As they get separated by their masses, proteins can be visualized in the gel and the gel show protein of interest among others. Another feature to separate proteins is their isoelectric point. This point represents the pH level where the protein has zero net charges. The technique that uses this property to separate proteins is called isoelectric focusing. When proteins go through a pH gradient gel, they will stop at the point where they have no net charge and get separated from the proteins with distinct isoelectric points. To get more specific results, isoelectric focusing can be coupled with SDS-PAGE (sodium dodecyl sulfate-polyacrylamide gel electrophoresis) in a technique called two-dimensional electrophoresis. In this technique, first isoelectric focusing is done as the sample goes through the gel horizontally and after proteins stop at their respective pH levels, vertical electrophoresis starts. So, the sample is separated according to the isoelectric points horizontally and by their masses vertically. The two-dimensional separation technique is employed to distinguish differences in two different states. Actually, distinct spots may be characterized by MALDI-TOF mass spectrometry.

2.1 Protein structure determination and screening databases

The structure and function of a protein are essential to characterize the linkage associated between common motifs and biochemical activity [34]. These features are normally determined by NMR or X-ray crystallography techniques [35]. NMR determines dynamic structure but the technique is limited to protein molecular mass. However, an instant picture of the protein structure with a relatively higher mass can be taken by X-ray crystallography. After elucidation of information from these structure determination techniques, scientists concluded that similar sequences show similar structural patterns [36]. Then, different databases which can display protein 2 and 3-dimensional structures have been developed. Determination of protein structure is important for not only understanding the function but also important for protein experiments, such as protein purification. Proteins have alpha helixes, turns, and beta sheets as secondary structures. Beta sheet structures and the outer surface of alpha helixes of proteins can accumulate within the cellular medium or stick to each other and other proteins during aggregation. *In vitro* experiments of proteins showed that proteins do not act like they are within the cell in terms of their stability, charge, and interactions properties. To understand the optimum conditions for protein purification, structure determination databases and pI calculation play crucial roles. In the case of His-tagged proteins, pI is one of the most important parameters for purification. pI is determined with a special calculation by considering amino acid sequences. NCBI Protein Data Bank provides the amino acid sequence of desired proteins (https://www.rcsb.org/). Several properties of the selected protein can be calculated by Expasy (Expert Protein Analysis System) Tool which is one of the most convenient tools on the web (https://web.expasy.org/compute_pi/) [36].

2.2 Swiss modeling and I-Tasser

Protein structure can be determined by bioinformatic tools such as Swiss Modeling or I-Tasser. Swiss model and I-Tasser are Protein Data Bank (PDB) dependent protein homology modeling databases. They use the known templates from PDB. Swiss Modeling is quicker than I-Tasser, however, I-Tasser produces better and more stable results. Swiss Modeling uses known sequences on the internet and generates data by comparing known structures. Both Swiss Modeling and I-Tasser have the advantage of understanding the main structure of the protein. Protein structure screening is a key factor to understanding interactions of proteins within themselves and with the environment. 3D modeled protein structures can be screened with commercially available tools like YASARA and Discovery Studio Tool. In these tools, not only protein structure is screened, but also domains of proteins can be separated, deleted and water molecules can be removed. UCSF Chimera and Autodock Tools also can be used for screening. After modeling, structures may be downloaded as PDB format and can be visualized by several programs: Discovery Studio, Autodock Vina, UCSF Chimera, etc. Interactions within the protein and secondary-tertiary structures can be obtained from these tools. All these tools work with a PDB file. There are other tools for protein structure determination and can be found at https://www.click2drug.org/. This website provides database links for distinct applications.

3. Importance of databases and protein structure determination in recombinant protein purification

Protein databases play a crucial role in bioinformatics and help to find information related to their research. In this way, all biological information becomes accessible through data mining tools saving time and resources. The first step in the study of a new protein is searching databases. Without the prior knowledge from such searches, previously known protein information could be missed, or an experiment could be repeated unnecessarily. There are hundreds of useful databases that can be used in protein research. However, in this study, the pI and 3D structure of the peptide were obtained using the Swiss database and Expasy Database. The purpose of the Swiss Database is to make protein structure modeling accessible. Therefore, with this database, the 3D shape of the protein provided us with information to understand the structure of the peptide and its interactions. Additionally, Expasy Database provides information about proteomics, post-translational modification prediction, primary, secondary, and tertiary

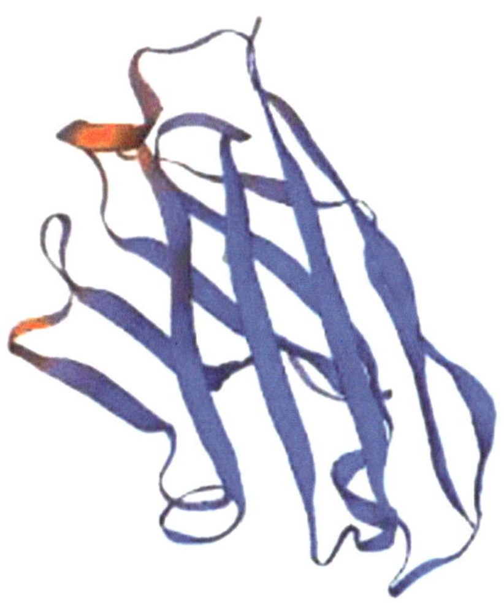

Figure 1.
Predicted structure of Sb#14. Sb#14 is a recombinant synthetic monoclonal antibody 14 used to detect spike protein of COVID-19 and used for immunodetection of the virus. SB#14 is modeled by the Swiss model to design purification steps for the TUSEB project.

structure analysis, sequence alignment, and pI of the protein [37]. The pI of the peptide provides the pH range where that peptide has a negative charge and prepares the buffer solution accordingly. Consequently, databases have key roles in biological research, and enormous data for protein structures, functions, and sequences can be generated by these available databases. These data offer essential information about our protein research as well. **Figure 1** provides the predicted structure for our research. SB#14 model indicates that the protein is formed from β-sheet structures.

4. Structure- and design-based difficulties in recombinant protein purification

4.1 Protein insolubility

Protein solubility is one of the most important protein properties and it can be defined as the protein concentration in a saturated solution that is in equilibrium with a solid phase [38]. Not only some extrinsic factors, including pH, ionic strength, temperature, and some solvent additives, can affect the protein solubility but also several intrinsic factors influence protein solubility. Moreover, the amino acids on the protein surface are the primary intrinsic factors that impact protein solubility [39]. Several studies have revealed the relationship between protein solubility and sequence-derived characteristics. Wilkinson and Harrison et al. provided a simple approach for predicting protein solubility from the sequence, which was further refined by Davis et al. [40]. The average charge, which is derived by the relative quantities of Asp, Glu, Lys, and Arg residues, and the concentration of turn-forming residues are the two parameters used in their solubility model (Asn, Gly, Pro, and Ser). In addition, Christendat et al. have demonstrated that insoluble proteins had more hydrophobic stretches (more than 20 amino acids), less glutamine (Q 4%), fewer negatively charged residues (DE 17%), and a higher percentage of aromatic amino acids (FYW >7.5%) than soluble proteins [41]. The affinity tag (His/GST) in recombinant protein purified by affinity chromatography allows the protein to be purified. However, affinity tag has been observed to alter the biological activity of the protein. Because a minor difference affects protein solubility, the choice of affinity tag at the N- or C-terminus is important when expressing a protein domain. Klock and colleagues investigated a nested collection of 2143N- and C-terminal truncations from 96 targets and found significant variance in both solubility and aggregation processes by changing just a few amino acids in a protein length [42]. Therefore, it is essential to analyze which end of the protein is hidden. Furthermore, if the three-dimensional structure is known, the tag should be kept in a solvent-accessible end. In this way, the solubility of the protein can be increased. Insoluble proteins can aggregate during the expression process. That is why the different parameters should be optimized. On the other hand, during downstream purification steps, protein aggregation can occur. In these cases, developing a suitable and optimized purification procedure for each protein is critical. Sb#14 hydrophobic nature (**Figure 1**) leads to solubility problems as well as aggregation. The protein sticks to larger proteins and this led to difficulties in purification.

4.1.1 Effects of imidazole on protein solubility

Imidazole is one of the most widely used organic compounds in protein affinity purification processes. It is used as a competitive agent to elute the histidine-tagged proteins.

High concentrated imidazole that includes protein samples should be eliminated after eluting from the nickel column by dialysis [43]. In spite of all precautions, his tagged Sb#14 sticks to other proteins and has low solubility, therefore, SEC is used for protein purification rather than affinity purification. As mentioned, SEC separates proteins according to the molecular weight of the molecule. SEC performed with Superdex 75 size-exclusion column 10/30 (GE Healthcare, Princeton, NJ, USA). Moreover, SEC (15 cm length with r, 3 cm column) used in this study separates aggregates readily. Lower molecular weight of Sb#14 provides an advantage in the purification process, recalling that larger proteins elute first. This custom SEC column was unique as the resin resolution is high while the column length is relatively lower. The choice of resin and column size helped to resolve Sb#14 from bacterial lysate in a single purification step. The peptide (MW: 12.468 g/mol, pI: 8.91) is small and prone to aggregate. Therefore, the single-step purification blocks the self-cleavage of protein domains.

4.1.2 Protein folding

The stability of the protein in various buffer compositions and pH levels with and without ligands should be determined. There are some useful websites for fold recognition that can be used to predict the protein fold (PSI-BLAST and SEARCH). Some proteins are misfolded and require the addition of a cofactor, or ligand to restore proper folding and increase stability. For instance, beta-sheets are more prone to form amyloid-like aggregates if there are other binding partners that support protein stabilization and folding [44]. If the protein has a large number of beta-sheets, aggregation may be observed. This can be explained by the tendency of sticking together at Sb#14 and leading to the formation of insoluble aggregates. Tris–HCl buffer is used to stabilize Sb# 14.

4.1.3 Reducing agents

To reduce aggregation, reducing agents such as dithiothreitol (DTT) may be used and added to the buffer. DTT is called Cleland's reagent and is used for protein reduction. However, a high concentration of DTT can reduce the nickel ion in the resin of the column. That is why the determination of the optimal concentration of DTT is essential. β-ME (Beta-mercaptoethanol) cleaves protein disulfide bonds (cystine), and TCEP (Tris phosphine hydrochloride) can also be used as reducing agents, considering longer half-life β-ME. DTT reacts easily with nickel ions whereas β-ME reacts easily with cobalt, copper ions, and other phosphate buffers [44]. A precaution is required to obtain optimal conditions.

4.1.4 Isoelectric point (pI) and pH

Each protein has a pI, where the protein's net charge is zero. Protein does not migrate at that point, and aggregation occurs [45]. On one hand, acidic proteins are likely to crystallize 0–2.5 pH units above their isoelectric point. On the other hand, basic proteins are more likely to crystallize 1.5–3 pH units below their pI. Hence, different pH values affect the protein's stability and solubility [46]. The pI of the peptide is important for us to know the pH range where that peptide has a negative charge and to prepare the buffer solution accordingly. That is why the pH of the buffer component is one of the most critical parameters. Sb #14 has a pI value of 8.91. This value set the pH parameter (pH:7.91) of the buffer used in the purification process.

4.2 Importance of protein isoelectric point in tagged protein purification

pI represents the pH level of a molecule where the net charge is zero. Amino acid composition of the protein can be used to calculate an estimated value with the help of databases [47]. If the pI is lower than the pH of a solution, protein will have a negative charge but if it is the opposite then the protein will have a positive charge. This feature can be used for purification purposes since it is a specific physicochemical parameter to distinguish between amphoteric molecules [39]. Also, it can be used to understand how solution pH can affect the protein stability in the pH range. So, buffers are used to keep proteins stable. To create an environment for protein to be stable, generally, the buffer is selected to have a pH level around the pI of the protein. If this difference between the pI of protein and pH of the solution gets larger then protein gets a greater net charge too. With this greater net charge, ionic compounds will be able to bind residues [48]. To avoid this unspecific interaction, the buffer's pH range should be selected accordingly to the protein's pI. And this knowledge of pH values with their effects on the proteins can be useful in the purification process. In tagged protein purification, affinity chromatography is a commonly used technique. The pH levels also affect this technique since affinity resins have their pH ranges to provide more stable links for not only the ligand and the bead but also for the tag and the ligand. While making decisions about the purification protocol, the affinity resins' and the tags' working pH ranges should be kept in mind to create a better environment and more stable interaction. Also choosing affinity resins and tags that have a wider range of pH that they can work may be useful for the purification of proteins.

4.3 Protein aggregation and importance of ionic strength

Proteins are special structures that work with covalent and non-covalent interactions. They have cellular wide roles, including signaling, structural and metabolic processes. Their special structural features and 3D architecture determine their roles and interactions. These forms of proteins are determined by the amino acid sequences [49]. Proteins are not synthesized in their functional form. When their translation process is finished, the primary protein structure is formed. After that, they form alpha helixes and beta sheets by hydrogen bonds. Alpha helixes and beta sheets interact with each other with weak interactions and disulfide bonds and tertiary structure is formed. In some instances, the quaternary structure may be formed when tertiary structures interact and eventually in all cases functional protein forms [55]. However, in some cases, proteins can accumulate and form aggregates which may cause failure in protein purification experiments. Mostly, beta-sheets tend to interact with each other and accumulate. This event can be exampled by amyloid aggregates in Alzheimer's disease. Recombinant protein aggregates resulted in the prevention of exposure of tags in tagged protein that causes failure in purification. Also, solubility prevents aggregate formation in proteins [50]. Protein aggregation can be prevented by adding salt to the proteins. Salt ions interact with the charged protein surface areas and prevent non-specific interactions, aggregation, and lower protein–protein interaction. However, a precaution is a must when preparing protein for binding experiments. Please note that high ionic concentration blocks ligand/protein binding experiments. As shown in **Figure 2**, Sb #14 is prone to aggregation and the process may be prevented/decreased through proper conditions.

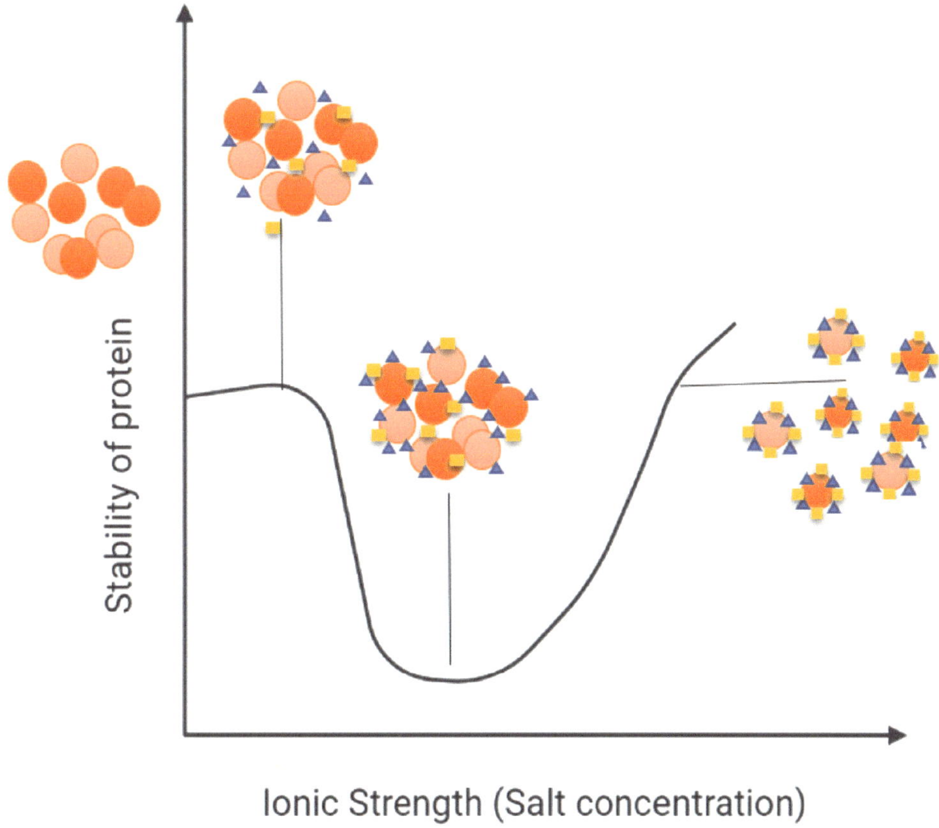

Figure 2.
Ionic strength is important for the separation of proteins from each other that can be aggregated. The strength of ions resists accumulation by preventing protein-protein interaction.

4.4 Usage of additional agents to prevent protein accumulation, attachment, and insolubility

Urea dissolves the aggregated protein solutions. The efficiency of the process is increased by taking the necessary purification steps [1, 51]. Among these processes, protein dissolving and refolding steps constitute the most important steps for optimal protein activity and higher recovery. The protein precipitate is generally separated from other cellular components by low-speed centrifugation after cell lysis. Because protein aggregates are denser than cellular components, the lysate proteins are precipitated by centrifugation and then dissolved using detergents such as urea, guanidine-HCl, high concentrations of chaotropic denaturants, sodium N-lauroyl sarcosine, SDS, N-acetyl trimethyl ammonium chloride [52]. Further, additional reducing agents such as DTT, cysteine, Triton X-100, β-ME are used to dissolve inclusion bodies. These agents retain cysteine residues, minimizing the formation of false and unnatural disulfide bonds in the protein solution. Metal-containing oxidation of cysteine is prevented by using chelating agents such as EDTA in dissolution buffers [44, 52]. By removing the soluble protein content, removing the chaotropic reagents, and diluting them directly into the renaturation buffer, the recombinant proteins are folded back into their native form [44]. Protein collapse is a higher-order reaction while protein folding is a lower-order reaction. Therefore, the aggregation rate is higher than the folding rate. Due to the kinetic competition that occurs, the increase in protein concentration decreases the folding efficiency of the protein. For accurate and efficient folding kinetics, the preferred protein concentration is used in the range of 10–50 μg.ml^{-1} [1, 53]. As explained in the section of 'Disulfide bond formation', recombinant proteins with multiple disulfide bonds in their structure tend to be in a correct folding process in the presence of both oxidizing and reducing agents for the formation of these bonds. The simplest way for oxidation is to oxidize the protein with air in the presence of a metal catalyst. Another common oxidation option is the addition of thiol agents containing compounds such as glutathione, cysteine, cysteamine to the protein mixture. The most commonly used thiol reagents are reduced/oxidized glutathione (GSH-GSSH), cysteine/cystine, DTT/GSSH, cysteamine compounds [1, 36]. There are also low-molecular-weight additives that help refolding process. There are studies on the use of additives such as acetone, DMSO, short-chain alcohols, PEG in the bioactive protein process. In addition, it has been observed that L-arginine/HCl reduces aggregation on protein. The 0.4–1 M arginine used in the studies also increases the protein folding efficiency by reducing the aggregation in the recombinant protein solution. This feature of arginine has been attributed to the interaction of the guanidino structure in its structure with tryptophan residues in proteins [44, 52–55]. Sb #14 was also treated with DTT but when overexpressed, the protein has solubility problems. The structure is highly prone to aggregation and solubility may be increased upon co-expressing chaperones/Heat Shock Proteins or yeast systems seem proper for preventing aggregation. Yet, this may not solve the problem but mutational studies may provide more soluble and stable structures.

5. Conclusion

Protein purification depends on several factors: resin type, solvent, ionic strength, pH, protein structural tendency to aggregation, buffer systems, protein structure, ligand if any, column dimension. For each factor, problems may be encountered. To eliminate these problems and decide on protein purification protocol, protein structural properties must be examined initially. Tandem purification steps may also increase the purification yield. However, self-cleavage of certain proteins or oxidation that may distort the protein function leads to problems. Therefore, several distinct protocols may be tested before purifying the targeted protein with high efficiency and functionality. Sb#14 structure mainly consists of β-sheets and overexpressing this petit protein lead aggregation. Solubility is another problem in the cellular milieu as Sb#14 hydrophobic nature interacts with other proteins in the lysate. Therefore, proper solvent selection (phosphate buffer) and adjusting the pH (1 unit lower than pI, 7.91) provide soluble protein. Further, we take advantage of the protein's lower molecular weight and employed a convenient resin (Superdex 75-separates 3000–70,000 molecular weights, most of the lysate elutes before Sb#14) and custom size column (15 cm length with r: 3 cm-lower pressure yet increase resolution). The purification was performed with high yield by AKTA go FPLC system. Additionally, co-expressing heat shock proteins with this type of protein may help in folding and dissolving aggregates. All these conditions must be tested for individual proteins for optimum purification yield.

Acknowledgements

Prof. Dr. Yusuf TUTAR acknowledges grant from TUSEB (Project # 8970-220-CV-01) and infrastructure grant from the University of Health Sciences-Turkey (Project #2017-041). Sb#14 is from Addgene (#153522).

Author details

Kubra Acikalin Coskun[1], Nazlıcan Yurekli[2], Elif Cansu Abay[2], Merve Tutar[3], Mervenur Al[4] and Yusuf Tutar[5,6,7]*

1 Faculty of Medicine, Division of Medicinal Biology, Department of Basic Sciences, Istanbul Aydin University, Istanbul, Turkey

2 Department of Molecular Biology and Genetics, Arel University, Istanbul, Turkey

3 Department of Molecular Biology and Genetics, Bilkent University, Ankara, Turkey

4 Hamidiye Faculty of Medicine, Division of Medicinal Biochemistry, Department of Basic Sciences, University of Health Sciences-Turkey, Istanbul, Turkey

5 Hamidiye Faculty of Pharmacy, Division of Biochemistry, Department of Basic Pharmaceutical Sciences, University of Health Sciences-Turkey, Istanbul, Turkey

6 Division of Molecular Oncology, Hamidiye Health Sciences Institutes, University of Health Sciences-Turkey, Istanbul, Turkey

7 Validebağ Research Center, University of Health Sciences-Turkey, Istanbul, Turkey

*Address all correspondence to: yusuf.tutar@sbu.edu.tr

References

[1] Singh SM, Panda AK. Solubilization and refolding of bacterial inclusion body proteins. Journal of Bioscience and Bioengineering. 2005;**99**(4):303-310. DOI: 10.1263/jbb.99.303

[2] Stollar EJ, Smith DP. Uncovering protein structure. Essays in Biochemistry. 2020;**64**(4):649-680. DOI: 10.1042/EBC20190042

[3] Buccitelli C, Selbach M. mRNAs, proteins and the emerging principles of gene expression control. Nature Reviews Genetics. 2020;**21**(10):630-644. DOI: 10.1038/s41576-020-0258-4

[4] Baneyx F. Recombinant protein expression in Escherichia coli. Current Opinion in Biotechnology. 1999;**10**(5):411-421. DOI: 10.1016/S0958-1669(99)00003-8

[5] Andersen DC, Andersen DC, and Krummen L. Recombinant protein expression for therapeutic applications Recombinant protein expression for therapeutic applications. Current Opinion in Biotechnology. May 2002;**1669**:117-123

[6] Rosano GL and Ceccarelli EA. Recombinant protein expression in Escherichia coli: Advances and challenges. Frontiers in Microbiology. 2014;**5**:1-17. DOI: 10.3389/fmicb.2014.00172

[7] Peleg Y, Prabahar V, Bednarczyk D, Unger T. Heterologous Gene Expression in E.coli. Current Opinion in Biotechnology. 2017;**1586**:33-43. DOI: 10.1007/978-1-4939-6887-9

[8] Schumann W, Ferreira LCS. Production of recombinant proteins in Escherichia coli. Genetics and Molecular Biology. 2004;**27**(3):442-453. DOI: 10.1590/S1415-47572004000300022

[9] de Marco A. Recombinant expression of nanobodies and nanobody-derived immunoreagents. Protein Expression and Purification. 2020;**172**:105645. DOI: 10.1016/j.pep.2020.105645

[10] Demain AL, Vaishnav P. Production of recombinant proteins by microbes and higher organisms. Biotechnology Advances. 2009;**27**(3):297-306. DOI: 10.1016/j.biotechadv.2009.01.008

[11] Sezonov G, Joseleau-Petit D, D'Ari R. Escherichia coli physiology in Luria-Bertani broth. Journal of Bacteriology. 2007;**189**(23):8746-8749. DOI: 10.1128/JB.01368-07

[12] Lee C, Kim J, Shin SG, Hwang S. Absolute and relative QPCR quantification of plasmid copy number in Escherichia coli. Journal of Biotechnology. 2006;**123**(3):273-280. DOI: 10.1016/j.jbiotec.2005.11.014

[13] Del Solar G, Espinosa M. Plasmid copy number control: An ever-growing story. Molecular Microbiology. 2002;**37**(3):492-500. DOI: 10.1046/j.1365-2958.2000.02005.x

[14] Browning DF, Godfrey RE, Richards KL, Robinson C, Busby SJW. Exploitation of the Escherichia coli lac operon promoter for controlled recombinant protein production. Biochemical Society Transactions. 2019;**47**(2):755-763. DOI: 10.1042/BST20190059

[15] Lalwani MA et al. Optogenetic control of the lac operon for bacterial chemical and protein production. Nature Chemical Biology. 2021;**17**(1):71-79. DOI: 10.1038/s41589-020-0639-1

[16] Peubez I et al. Esters of dicarboxylic acids as additives for lubricating oils.

Tribology International. 2006;**39**(6):560-564. DOI: 10.1016_j.triboint.2005.06.001

[17] Kimple ME, Brill AL, and Pasker RL. Overview of affinity tags for protein purification. Current Protocols in Protein Science. 2013;**73**:608-616. DOI: 10.1002/0471140864. ps0909s73

[18] Terpe K. Overview of tag protein fusions: From molecular and biochemical fundamentals to commercial systems. Applied Microbiology and Biotechnology. 2003;**60**(5):523-533. DOI: 10.1007/s00253-002-1158-6

[19] Walls D, Walker JM. Protein Chromatography. Protein Chromatography. 2017;**1485**:423. DOI: 10.1007/978-1-4939-6412-3

[20] Costa SJ, Coelho E, Franco L, Almeida A, Castro A, Domingues L. The Fh8 tag: A fusion partner for simple and cost-effective protein purification in Escherichia coli. Protein Expression and Purification. 2013;**92**(2):163-170. DOI: 10.1016/j.pep.2013.09.013

[21] Costa S, Almeida A, Castro A, and Domingues L. Fusion tags for protein solubility, purification, and immunogenicity in Escherichia coli: The novel Fh8 system. Frontiers in Microbiology. 2014;**5**(FEB):63. DOI: 10.3389/fmicb.2014.00063

[22] Chang HM, Yeh ETH. Sumo: From bench to bedside. Physiological Reviews. 2020;**100**(4):1599-1619. DOI: 10.1152/physrev.00025.2019

[23] Dumon-Seignovert L, Cariot G, Vuillard L. The toxicity of recombinant proteins in Escherichia coli: A comparison of overexpression in BL21(DE3), C41(DE3), and C43(DE3). Protein Expression and Purification. 2004;**37**(1):203-206. DOI: 10.1016/j.pep.2004.04.025

[24] Studier FW. Protein production by auto-induction in high density shaking cultures. Protein Expression and Purification. 2005;**41**(1):207-234. DOI: 10.1016/j.pep.2005.01.016

[25] Jäger VD et al. Catalytically-active inclusion bodies for biotechnology—General concepts, optimization, and application. Applied Microbiology and Biotechnology. 2020;**104**(17): 7313-7329. DOI: 10.1007/s00253-020-10760-3

[26] Messens J, Collet JF. Pathways of disulfide bond formation in Escherichia coli. International Journal of Biochemistry and Cell Biology. 2006;**38**(7):1050-1062. DOI: 10.1016/j.biocel.2005.12.011

[27] Thomas JG, Baneyx F. ClpB and HtpG facilitate de novo protein folding in stressed Escherichia coli cells. Molecular Microbiology. 2000;**36**(6):1360-1370. DOI: 10.1046/j.1365-2958.2000.01951.x

[28] Sørensen HP, Mortensen KK. Advanced genetic strategies for recombinant protein expression in Escherichia coli. Journal of Biotechnology. 2005;**115**(2):113-128. DOI: 10.1016/j.jbiotec.2004.08.004

[29] Campani G, Gonçalves da Silva G, Zangirolami TC, Perencin de Arruda Ribeiro M. Recombinant Escherichia coli cultivation in a pressurized airlift bioreactor: Assessment of the influence of temperature on oxygen transfer and uptake rates. Bioprocess and Biosystems Engineering. 2017;**40**(11):1621-1633. DOI: 10.1007/s00449-017-1818-7

[30] Ferrer M, Lünsdorf H, Chernikova TN, Yakimov M, Timmis KN, Golyshin PN. Functional consequences of single:Double ring transitions in chaperonins: Life in the cold. Molecular

Microbiology. 2004;**53**(1):167-182. DOI: 10.1111/j.1365-2958.2004.04077.x

[31] Wingfield PT. Overview of the purification of recombinant proteins. Current Protocols in Protein Science. 2015;**80**:6.1.1-6.1.35. DOI: 10.1002/0471140864.ps0601s80

[32] Berg JM, Tymoczko JL, Stryer L. Section 4.1, The purification of proteins is an essential first step in understanding their function. In: Biochemistry. 5th ed. New York: W H freeman; 2002. Available from: https://www.ncbi.nlm.nih.gov/books/NBK22410/

[33] Harcum S. Purification of protein solutions. Biologically Inspired Textiles. 2008;**1**:26-43. DOI: 10.1533/9781845695088.1.26

[34] Righetti PG. Electrophoresis| Isoelectric focusing. Encyclopedia of Analytical Science. 2005;**1079**:382-392. DOI: 10.1016/b0-12-369397-7/00124-2

[35] Tramontano A. Function Prediction. Oxfordshire, United Kingdom: Taylor & Francis Group; 2005. pp. 45-67. DOI: 10.1201/9781420035001.ch3

[36] Factors R. Chapter 11: Orthodontic Treatment of Class III Malocclusion. Vol. 1990. Amsterdam, Netherlands: Elsevier; 1999. p. 306. DOI: 10.1007/978-1-4939-7315-6

[37] Waterhouse A, Bertoni M, Bienert S, Studer G, Tauriello G, Gumienny R, et al. SWISS-MODEL: Homology modelling of protein structures and complexes. Nucleic Acids Research. 2018;**46**:W296-W303

[38] Gasteiger E, Gattiker A, Hoogland C, Ivanyi I, Appel RD, Bairoch A. ExPASy: The proteomics server for in-depth protein knowledge and analysis. Nucleic Acids Research. 2003;**31**(13):3784-3788. DOI: 10.1093/nar/gkg563. PMID: 12824418; PMCID: PMC168970.

[39] Kramer RM, Shende VR, Motl N, Pace CN, Scholtz JM. Toward a molecular understanding of protein solubility: Increased negative surface charge correlates with increased solubility. Biophysical Journal. 2012;**102**(8):1907-1915. DOI: 10.1016/j.bpj.2012.01.060

[40] Trevino SR, Scholtz JM, Pace CN. Amino acid contribution to protein solubility: Asp, Glu, and ser contribute more favorably than the other hydrophilic amino acids in RNase Sa. Journal of Molecular Biology. 2007;**366**(2):449-460. DOI: 10.1016/j.jmb.2006.10.026

[41] Davis GD, Elisee C, Mewham DM, Harrison RG. New fusion protein systems designed to give soluble expression in Escherichia coli. Biotechnology and Bioengineering. 1999;**65**(4):382-388. DOI: 10.1002/(SICI)1097-0290(19991120)65:4<382::AID-BIT2>3.0.CO;2-I

[42] Smialowski P, Martin-Galiano AJ, Mikolajka A, Girschick T, Holak TA, Frishman D. Protein solubility: Sequence based prediction and experimental verification. Bioinformatics. 2007;**23**(19):2536-2542. DOI: 10.1093/bioinformatics/btl623

[43] European Commission EUROSTAT. Glossary:Special-purpose entity (SPE). Nature Methods. 2008;**5**(2):135-146. DOI: 10.1038/nmeth.f.202.Protein

[44] Bhat E, Abdalla M, Rather I. Key Factors for Successful Protein Purification and Crystallization. Global Journal of Biotechnology and Biomaterial Science. 2018;**4**(1):001-007. DOI: 10.17352/gjbbs.000010

[45] Novák P, Havlíček V. Protein extraction and precipitation.

In: Proteomic Profiling and Analytical Chemistry: The Crossroads. Second ed. Amsterdam, Netherlands: Elsevier; 2016. pp. 52-62. DOI: 10.1016/B978-0-444-63688-1.00004-5

[46] Friedman DB, Hoving S, Westermeier R. Chapter 30 isoelectric focusing and two-dimensional gel electrophoresis. Methods in Enzymology. 2009:515-540. DOI: 10.1016/s0076-6879(09)63030-5

[47] Lee K. Protein Folding. Genome Biology. 2001;**2**(1):1-20. DOI: 10.1186/gb-spotlight-20010313-01

[48] Dobson C. Protein folding and misfolding. Nature. 2003;**426**:884-890. DOI: 10.1038/nature02261

[49] Roberts CJ. Protein aggregation and its impact on product quality. Current Opinion in Biotechnology. Dec 2014;**30**:211-217. DOI: 10.1016/j.copbio.2014.08.001. Epub 2014 Aug 28. PMID: 25173826; PMCID: PMC4266928

[50] Bao RM, Yang HM, Yu CM, Zhang WF, Tang JB. An efficient protocol to enhance the extracellular production of recombinant protein from Escherichia coli by the synergistic effects of sucrose, glycine, and triton X-100. Protein Expression and Purification. 2016;**126**:9-15. DOI: 10.1016/j.pep.2016.05.007

[51] Fahnert B. Using folding promoting agents in recombinant protein production: A review. Recombinant Gene Expression. 2012;**824**:3-36

[52] Cline DJ, Redding SE, Brohawn SG, Psathas JN, Schneider JP, Thorpe C. New water-soluble phosphines as reductants of peptide and protein disulfide bonds: Reactivity and membrane permeability. Biochemistry. 2004;**43**(48):15195-15203. DOI: 10.1021/bi048329a

[53] Kumar A et al. Optimization and efficient purification of recombinant Omp28 protein of brucella melitensis using triton X-100 and β-mercaptoethanol. Protein Expression and Purification. 2012;**83**(2):226-232. DOI: 10.1016/j.pep.2012.04.002

[54] Mayer M, Buchner J. Refolding of inclusion body proteins. Methods in Molecular Medicine. 2004;**94**:239-254. DOI: 10.1385/1-59259-679-7:239

[55] Gunner MR, Mao J, Song Y, Kim J. Factors influencing the energetics of electron and proton transfers in proteins. What can be learned from calculations. Biochimica Et Biophysica Acta (BBA) – Bioenergetics. 2006;**1757**(8):942-968. DOI: 10.1016/j.bbabio.2006.06.005

Chapter 2

Protein Detection in Clinical Diagnosis and Management of Prevalent Neurodegenerative Diseases and Metabolic Disorders

Ohanube A.K. Goodluck, Obeta M. Uchejeso and Ikeagwulonu R. Chinaza

Abstract

An accurate diagnosis gives leeway to cost-effective treatments. However, many diseases continue to evolve; hence, their etiology is sometimes missed due to the procedures used during diagnosis. Protein-related diseases include proteopathies (proteinopathies) such as neurodegenerative diseases and metabolic disorders like protein-energy malnutrition and some hormonopathies. Hormonopathies are associated with the change in the production of hormones. Diabetes mellitus, a type of hormonopathy, is reviewed in this work alongside neurodegenerative diseases and protein-energy malnutrition. This chapter aims to elucidate more on the diagnosis of these diseases considering the structure and function of their proteins viz-a-viz their deficiencies and hyper-production in man. Their pathogenesis and the principles underlying their diagnosis are further discussed to optimize the management of these diseases among patients.

Keywords: prion, proteinopathy, hormonopathy, marasmus, kwashiorkor, alzheimer, parkinson, huntington, diabetes, neurodegenerative disease

1. Introduction

Medical laboratory diagnosis has given hope to detect and efficiently treat or manage diseases, and protein-related diseases are not an exception. The diagnosis ranges from the least like urinalysis to the highly sophisticated methodologies involving molecular techniques such as polymerase chain reaction. This chapter discusses the neurodegenerative disease and the pathologic conditions associated with protein-derived hormones and protein-energy malnutrition, focusing on their diagnosis and management.

Proteinopathy (proteopathy) is a disease mainly characterized by the production of aberrant proteins. Here, there could be a defect in the structure or function of the proteins produced, which could also reflect in the over-secretion or under-secretion of these proteins. Neurodegeneration involves a gradual loss of neuronal structure or function, invariably leading to cellular death [1]. Neurodegenerative proteinopathies (proteopathies) are neurodegenerative diseases that have abnormally produced proteins that could result from the changes in the structure and function of these proteins. Their pathological examination reveals their link with aberrant proteins. Their prevalence in the United States of America is shown in **Table 1**. Popular

Disease	Age-adjusted deaths per 100,000
Prion disease	0.1319
Alzheimer's disease	233.8
Parkinson's disease	65.3
Frontotemporal dementia	66.7
Amyotrophic lateral sclerosis	5

Table 1.
Prevalence of some common neurodegenerative diseases in the United States [1–5].

Disease	Host	Route of exposure
Kuru	Fore people of New Guinea	Ritualistic cannibalism
CJD:		
Iatrogenic	Human	Invasive hospital equipment, implants, transplant organs, and tissues
Familial	Human	Germline mutations in the *PrP* gene
Sporadic	Human	Somatic mutation or spontaneous conversion of PrP^c into PrP^{sc}
Variant CJD		Ingestion of bovine prions
Gerstmann–Straussler–Scheinker syndrome	Humans	Germline mutations in *PrP* gene
Fatal familial insomnia	Humans	Germline mutation in the *PrP* gene (D178N, M129)
Sporadic fatal insomnia	Humans	Somatic mutation or spontaneous conversion of PrP^c into PrP^{sc}
Scrapie	Sheep	Genetic mutation in sheeps
Bovine spongiform encephalopathy	Cattle	Infection by prion contaminated meat and bone meals
Transmissible mink encephalopathy	Mink	Infection with prions from sheep or cattle
Chronic wasting disease	Mule, deer, elk	Unknown
Feline spongiform encephalopathy	Cats	Infection with beef contaminated with prions
Exotic ungulate encephalopathy	Greater Kudu, Nyala, Oryx	Infection with prion contaminated bone meal and meat

PrP, prion protein; PrP^c, cellular prion protein; PrP^{sc}, scrapie isoform of the prion protein; CJD, Creutzfeldt–Jacob disease.

Table 2.
Pathogenic feature of the types of prion disease [1, 2].

Disease	Protein	Histopathological identification
Prion disease [1, 2]	PrP^{Sc}	PrP^{sc} amyloid plaques
Alzheimer's disease [6, 7]	Aβ	Aβ amyloid plaques
	Tau	Paired helical filaments in neurofibrillary tangles
Parkinson's disease [8–10]	ά-Synuclein	Lewy bodies
Frontotemporal dementia [1, 2]	Tau	Straight filaments and paired helical filaments
Pick's disease [1, 2]	Tau	Pick bodies
Amyotrophic lateral sclerosis [1, 2]	Neurofilament	Neuronal aggregates
Huntington's disease [1, 2, 11, 12]	Huntington	Nuclear inclusions of CAG repeats
Spinocerebellar ataxia [1, 2]		
Type 1	Ataxin 1	Nuclear inclusions
Type 2	Ataxin 2	Cytoplasmic inclusions
Machado–Joseph disease	Ataxin 3	Nuclear inclusions

Table 3.
Protein detection in prevalent neurodegenerative diseases.

examples of neurodegenerative diseases are amyotrophic lateral sclerosis, Parkinson's disease, Alzheimer's disease, and Huntington's disease, which may be grouped as non-infectious neurodegenerative proteinopathies, and prion diseases considered infectious neurodegenerative proteinopathies [1, 2]. The pathogenic feature of the variety of prion disease, an infectious neurodegenerative proteinopathy, is shown in **Table 2**, while protein deposition seen in most common neurodegenerative diseases is shown in **Table 3**.

2. Infectious neurodegenerative proteinopathies

2.1 Prion diseases

Infectious neurodegenerative proteinopathies are proteinopathies that are communicable diseases, and their characteristic symptom is neurodegeneration in humans. Prion disease is an excellent example of such a disease. Some types of prion disease, also known as transmissible spongiform encephalopathies (TSEs), such as Creutzfeldt–Jakob disease (CJD), Kuru, Gerstmann–Straussler–Scheinker syndrome (GSS), and fatal familial insomnia, are a group of fatal neurodegenerative diseases. They are incurable but manageable; they can affect humans and animals and are sometimes transmitted to humans by infected meat products. The clinical progression is over weeks, progressing to akinetic mutism with a median disease duration of 20 weeks. Prodromal features, present in around 30% of cases, include depression, fatigue, weight loss, headaches, insomnia, general malaise, and ill-defined pain sensations. Additionally, we see myoclonus and mental deterioration, while neurological features include pyramidal signs, extrapyramidal signs, cerebellar ataxia, and cortical blindness [13–15]. The pathogenic features of the prion disease are shown in **Table 2**. Risk factors for prion disease are positive family history, eating meat infected by "mad cow disease," infection from receiving contaminated organs or tissues including corneal tissue, or contaminated medical equipment [16].

The central feature of the pathogenesis seen in most types of prion disease is the post-translational conversion of host-encoded, normal, healthy, cellular prion protein (PrP^C) to an abnormal infectious isoform, termed scrapie isoform of the prion protein (PrP^{Sc}) or (PrP^{res}), which is an alternatively folded variant of the cellular prion protein, PrP^C [15–17]. This misfolding of PrP^{Sc} is possible due to the more important content of the β-sheet structure, which aggregates to form medium and large-size polymers [18]. Studies have proposed different functions of PrP^C, such as the roles in apoptosis, neuroprotection, oxidative stress, transmembrane signaling, cell adhesion, myelination, and trafficking of metal ions. The critical event in the stages of prion disease is the structural and conformational change of PrP^C to the disease-associated misfolded form, PrP^{Sc} [18]. This conversion changes PrP^C from a protein characterized by alpha-helices to a partially protease-resistant misfolded protein filled with beta-sheets (β-sheets). Proteinase K (PK) partially digests PrP^{Sc} and is often used to determine the presence of misfolded PrP^{Sc}. PrP^{Sc} accumulates in different brain regions as distinct types of deposits depending on the animal species and strains of the infectious agent. The incapacitation of the critical biological function of PrP^C is one possible mechanism by which PrP^{Sc} formation might lead to degeneration of neurons. Another possible mechanism by which PrP^{Sc} formation might be linked to the disease is by direct toxicity of the misfolded protein [15–18]. **Table 2** shows the list of prion diseases and their routes of infection. Prion disease affecting animals has been included for academic purposes; however, our discussion is focused on those affecting humans.

2.2 Etiology and clinical manifestation

Human prion diseases can be grouped etiologically into sporadic, inherited, and acquired forms [13]. The following paragraphs shall consider the different etiological classifications of prion diseases, stating the respective examples.

2.2.1 Sporadic cases

More than 80% of the occurrence is sporadic cases of human prion disease, which presents as Creutzfeldt–Jakob disease (sporadic CJD). The cause of sporadic CJD is unknown, although it is hypothesized to include somatic mutation of the prion protein gene (*PRNP*) or the spontaneous conversion to PrP^{Sc} form of PrP^C. A polymorphism that occurs at residue 129 of human PrP (encoding either methionine (M) or valine (V)) strongly influences the susceptibility to human prion diseases. About a third of Europeans are homozygous for the more frequent methionine allele, half are heterozygous, and a tenth is homozygous for valine. Homozygosity at *PRNP* codon 129 is a causal factor to the development of sporadic and acquired CJD. Polymorphic homozygosity favors the occurrence of most sporadic CJD. This susceptibility factor is also vital in the inherited forms of CJD, most especially in vCJD. All hospitalized cases studied so far have been homozygous for codon 129 methionine of *PRNP*. Additionally, a haplotype for *PRNP* susceptibility has been identified, indicating additional genetic susceptibility to sporadic CJD at or near the *PRNP* locus [15–18].

Classical sporadic CJD presents with a rapidly progressive multifocal dementia predominantly with myoclonus. The onset is usually in the 45–75 years age group, with the median age at death of 68 years. The clinical progression expends over weeks, progressing to akinetic mutism with a median disease duration of 5 months. Prodromal features, present in about 30% of cases, include fatigue, insomnia, headaches, weight loss, depression, malaise, and non-specific pain sensations. In addition to mental deterioration and myoclonus with cerebellar ataxia, frequent additional neurological features include extrapyramidal signs, pyramidal signs, and cortical blindness [13, 18].

Atypical forms of sporadic CJD are seen in about 10% of cases of CJD, and they have a longer duration of a clinical course spanning over 2 years. Here, cerebellar ataxia is seen instead of cognitive impairment. Hence, it is termed ataxic CJD [19]. Heidenhain's variant of CJD refers to conditions in which cortical blindness is marked with severe involvement of the occipital lobes. The panencephalopathic type of CJD is more common in Japan; it presents with extensive degeneration of the cerebral white matter and spongiform vacuolation of the gray matter [20].

2.2.2 Inherited prion disease

About a fifth of human prion diseases is associated with autosomal dominant pathogenic mutations in *PRNP* [21–23]. The mechanism by which pathogenic mutations in *PRNP* cause prion disease is yet to be elucidated; however, it is believed that in most cases, it involves a mutation that leads to an increased tendency of PrP^{C} to form PrP^{Sc}. Even though pieces of evidence abound in congruence with this, this may partly be related to the decreased thermodynamic stability of mutated PrP^{C} [24, 25].

Traditionally, inherited prion diseases have been classified by the presenting clinical syndrome, falling into three main sub-divisions: GSS, CJD, or FFI. GSS is seen in people in their 40s; it classically presents as chronic cerebellar ataxia with pyramidal features with dementia seen much later in a clinical course that is usually longer than in classical CJD [13, 23]. Fatal familial insomnia (FFI) has its pathognomonic feature as progressive chronic insomnia, dementia and dysautonomia, selective thalamic degeneration, and is mainly associated with a missense mutation at codon 178 of *PRNP* (3); its sporadic form with no causative mutation in *PRNP* have been reported [13, 23, 24]. Another form of inherited prion disease, though extremely rare, is variably protease-sensitive prionopathy (VPSPr). VPSPr is similar to CJD; however, the protein is less sensitive to digestion. It is more likely to affect people in their seventh decade of life with a family history of dementia. The existence of phenotypic overlap between individuals with different mutations and even in family members with the same *PRNP* mutation indicates that accurate classification of inherited human prion diseases should be based upon mutation alone [24–26]. Due to the extensive phenotypic variability associated with inherited prion disease and its ability to mimic other neurodegenerative conditions, notably Alzheimer's disease, *PRNP* analysis should be considered in all patients with undiagnosed dementing ataxic disorders [13, 23, 26].

2.2.3 Acquired prion disease

Human prion diseases are transmissible diseases; their acquired forms have, however, until recently, been confined to rare and unusual situations. They include the iatrogenic CJD, Kuru, and variant CJD.

The two most prevalent causes of iatrogenic CJD occurring through the medical procedure are the implantation of grafts of dura mater and treatment with growth hormone derived from the pituitary glands of human cadavers [13]. Less frequent causes of human prion disease have been associated with the iatrogenic transmission of CJD during corneal transplantation, infected electroencephalographic (EEG) electrode implantation, and surgical operations using contaminated instruments or apparatus [13, 27, 28]. The clinical presentation in iatrogenic forms of human prion disease appears to be related to their etiology and, in particular, the route of exposure to human prions [13]. Peripheral routes of infection are commonly associated with more extended incubation periods and usually present with a Kuru-like syndrome, in which ataxia is common, while dementia is rare at the onset. Conversely, patients with dura mater graft-related exposure to human prions, in which infectivity is placed proximal to the brain, usually have a clinical presentation that looks like sporadic CJD, although exceptions with unusual clinical features have been reported [13, 27–29].

Kuru is a disease that used to be predominant among cannibals in the Fore tribe of the Eastern Highlands in Papua New Guinea, but it is now rare due to consistent enlightenment and rules that abolished such culture [29, 30]. It is caused by eating prion ladened human brain tissue. The central clinical feature of Kuru is progressive cerebellar ataxia, and in sharp contrast to sporadic CJD, dementia is late and may be absent. A prodrome and three clinical stages consisting of an ambulatory stage, a sedentary stage, and a tertiary stage have been described [13, 23, 29]. Remarkably, Kuru demonstrates that incubation periods of infection with human prions can exceed 50 years [29]. The *PRNP* codon 129 genotype has been identified to have a pronounced effect on Kuru in terms of the incubation periods and susceptibility, and most elderly survivors of the kuru epidemic are heterozygotes [26, 30, 31]. The glaring survival advantage for codon 129 heterozygotes gives a cue for a robust basis for selection pressure in the Fore clan [13, 23, 26]. However, analyzing the global haplotype diversity and frequency of the alleles responsible for coding and non-coding polymorphisms of *PRNP*, an older and widely spread balancing selection at this locus has more unusual variation because of heterozygote advantage is suggestive [23, 29, 31]. Only a few human genes present evidence for balancing selection. With the biochemical and physical evidence of cannibalism on five continents, one explanation is that cannibalism resulted in prion disease epidemics in human prehistory, thus imposing balancing selection on *PRNP* [23, 29, 31].

The variant CJD is an infectious type of disease that is related to "mad cow disease." Eating meat that has been inflicted with bovine spongiform encephalopathy (BSE) may cause the disease in humans, as seen in the United Kingdom years back [13, 27]. The meat may cause abnormal development of normal human prion protein. The disease is associated with iatrogenic conditions. This disease usually affects younger people and is rare in most developed nations [13, 27, 32].

2.3 Diagnosis of prion

Prion disease can be provisionally diagnosed using the clinical signs and symptoms presented alongside the taking of history. Neurologic and visual examinations could be done to ascertain nerve damage and vision loss. Prion diseases such as CJD can be diagnosed via MRI, PET, and CT scans of the brain and body; and spinal tapped cerebrospinal fluid (CSF). Electroencephalogram, which analyses brain waves, could also be used; this painless test requires placing electrodes on the scalp. At the same time, some centers choose to do blood and urine tests, which involves immunologically based analysis. Raised cerebrospinal fluid 14-3-3 protein, S-100, and neuronal-specific enolase (NSE), although unspecific for CJD, may be helpful diagnostically in the appropriate clinical context [13].

Prions lack DNA or RNA, so PCR or other nucleic acid-based tests cannot identify them. Hence, the strategy is to mix the test material with the proteinase K (PK) enzyme, which digests the regular portion of prion protein but cannot digest any of the portions, which appears abnormal. Some other techniques aim at detecting the residual protein (PrP^{Sc}) after digestion. Methods relying on PK digestion are less sensitive than those that do not rely on it because the former reduces the small amount of original PrP^{Sc} captured [33].

The most sensitive, crucial, precise, but uncommon immunoassay method of confirmatory diagnosis is by identifying the disease-causing PrP isoform (PrP^{Sc}) using the conformation-dependent immunoassay (CDI) laboratory method [15, 34, 35]. The CDI is the only immunoassay that measures both the protease-resistant and protease-sensitive forms of PrP^{Sc} [14]. The CDI was developed to quantify PrP^{Sc} in tissue samples from mammals producing prions. Sandwich CDI represents a rapid, robust, powerful tool to study prions in bodily fluids of CJD/vCJD patients, with a turnaround time of 12–24 hours [15, 34]. Safar et al., in their experiment, showed the superior performance of the CDI in diagnosing prion disease compared to the routine neuropathologic examination and immunohistochemistry (IHC). Hence, they proposed using CDI in place of these earlier mentioned methods [14, 33].

2.4 Managing prion diseases

Prion diseases rarely have a cure; hence they are managed using certain medications, which could slow their progress. This management focuses on keeping people with these diseases as safe and comfortable as possible despite progressive and debilitating symptoms.

Effective anti-prion agents may have broader implications due to the adverse effects associated with them. Several therapeutic approaches include polyanionic, polycyclic drugs such as pentosan polysulfate (PPS), which prevent the conversion of PrP^{c} to PrP^{res} and might also sequester and down-regulate the protease-resistant prion protein (PrP^{res}). Polyanionic compounds might also help to clear PrP^{res}. Treatments aimed at the laminin receptor, an essential accessory molecule in converting PrP^{c} to PrP^{res}—neuroprotection, immunotherapy, siRNA, and antisense approaches, have provided some experimental cues [28].

In drug development, the PrP^{C}, PrP^{Sc} (PrP^{res}), or the process of its conversion are the targets. Pentosan polysulphate (PPS) is presumed to act as a coreceptor for PrP on the cell surface in competition with endogenous heparin sulfate proteoglycans and shows the ability to inhibit the formation of new PrP^{Sc} in neuroblastoma cells. Quinacrine is thought to prevent PrP^{Sc} polymerization by stabilizing PrP^{C} and reducing its conversion to PrP^{Sc}. Doxycycline reverses the protease resistance of PrP^{Sc} extracted from CJD brains and prolongs the survival of animals experimentally infected with prions, even when given at the onset of clinical signs [33, 36–38]. Active and passive immunization are two significant aspects of immunotherapy. Resveratrol is an essential compound with antioxidant, anti-allergy, anti-aging, and neuroprotective activities, and it has been reported to eliminate prion replication *in vitro* and prion infection *in vivo*. The ubiquitin (Ub)-proteasome system (UPS) is the first line of defense in degrading soluble misfolded proteins. Conversion from PrP^{C} into PrP^{Sc} may involve chaperones and Ub ligases for UPS-dependent protein quality control. Enhanced UPS aims to stimulate the degradation of PrP^{Sc}. The autophagy-lysosome system is another quality control system to remove the misfolded proteins [36–38]. Studies have alluded that rapamycin can activate autophagy *in vitro* and delay disease onset in rodents with prion disease [39]. Autophagy could also lead to PrP^{Sc} clearance in cell models and prolong the lifespan of prion-infected mice [36].

3. Non-infectious neurodegenerative proteinopathies

3.1 Alzheimer disease

Alzheimer's disease (AD), a progressive neurodegenerative disorder, is the leading cause of dementia among geriatrics [6, 40]. It affects over 27 million persons worldwide, and prediction shows that over 86 million people would be affected by 2050 [7]. It is characterized by difficulty solving problems, memory loss, disorientation in time and space, among others [41]. This disease was first described in 1906 at a conference in Tubingen, Germany, by Alois Alzheimer [41]. Aging seen in the absence or presence of dementia of the Alzheimer type (DAT) is associated with loss of weight; hence, accelerated weight loss of idiopathic origin may herald the onset of DAT, aiding its clinical diagnosis [42]. The significant risk factors of this multifactorial disease include apolipoprotein E 4, hypercholesterolemia, genotype, traumatic brain injury, family history, age, obesity, hypertension, diabetes, and low level of education [6].

A complex array of molecular events has been implicated in the pathogenesis of AD. The major pathological characteristics of AD brains are senile, neurofibrillary tangles, plaques, and neuronal loss [6, 7, 40, 41]. The pathogenic mechanism implicated here seems elusive; however, oxidative stress has been identified as a leading factor in the initiation and progression of the ailment [43]. The excessive reactive oxygen species may be generated from mitochondria dysfunction and aberrant accumulation of transition metals, while the abnormal accumulation of amyloid-beta (Aβ) and tau proteins appears to promote the redox imbalance leading to neurotoxicity [41–43]. Additionally, oxidative stress may augment the production and aggregation of Aβ and facilitate the phosphorylation and polymerization of tau, leading to a vicious cycle that promotes the initiation and progression of AD [43]. Researches are gradually drifting from the simple assumption of the original amyloid hypothesis to new theories of pathogenesis, which include gamma oscillations, cerebral vasoconstriction, prion transmission, growth hormone secretagogue receptor 1α (GHSR1α)-mediated mechanism, and infection [44].

3.1.1 Diagnosis

The disease morphologically features an overall loss of synapses and neurons and an overall reduction in brain volume. The neuropathologic examination has been identified as the gold standard for diagnosing Alzheimer's disease (AD). However, popular opinion has it that histologic examination is the best indicator of AD diagnosis. Thus, an autopsy may gradually become the gold standard for determining clinical diagnostic accuracy rates [7, 45, 46]. A routine examination is better done with magnetic resonance (MR) or computed tomographic (CT) imaging. In the early onset of the disease, coronal MR images have been helpful to document or quantify the atrophy of both the hippocampus and entorhinal cortex. At the same time, subtraction and volumetric MR techniques can be used to quantify and monitor rates of regional atrophy and dementia progression. Positron emission tomography (PET) coupled with single-photon emission CT is helpful in the differential diagnosis of AD from other dementias associated with the cortical and subcortical dementias and may also be of prognostic value. Values from the MR are also used to monitor treatment effects in clinical trials of antidementia agents and cognitive enhancers [7, 40, 46].

Additionally, PET studies have shown that subtle abnormalities may occur at the prodromal stages of AD and in subjects bearing susceptibility genes. PET ligands may be of value in identifying amyloid plaques. Functional MR-based memory challenge tests are also beneficial [7, 40, 46].

Peripheral biomarkers are also beginning to gain ground in the diagnosis of Alzheimer's disease. This gives room for presymptomatic detection of disease, which could be valuable for monitoring the efficacy of disease interventions during clinical trials. CSF has long remained the sample of choice for biomarkers for many scientists until some Australian scientists developed theirs using blood. A biomarker panel that was about 85% sensitive and 93% specific was developed. The plasma markers in this biomarker panel that was significantly increased were cortisol, pancreatic polypeptide, β_2 microglobulin, insulin-like growth factor binding protein 2, and vascular cell adhesion molecule 1. There was also CD40, carcinoembryonic antigen, matrix metalloprotein 2, macrophage inflammatory protein 1α, superoxide dismutase, and homocysteine. In AD, these markers were decreased (apolipoprotein E, epidermal growth factor receptor, hemoglobin, calcium, zinc, interleukin 17, and albumin). This panel of plasma biomarkers was proven to be efficient, as it distinguished individuals with AD from cognitively healthy control subjects with high precision [7, 46].

Nevertheless, the prominent causal factors for AD development are genetic mutation involving genes encoding for proteins such as presenilin 1 (PSEN1), presenilin 2 (PSEN2), and amyloid precursor protein (APP). Usually, at an early age between the third and fifth decade of life, about half of the carriers of such mutations develop AD-type dementia. The hallmark of AD includes the accumulation of Aβ as senile plaques and aggregating hyperphosphorylated tau-mediated neurofibrillary tangles, NFTs for short [6, 7].

3.1.2 Management

Previous researches implicated an instability in the homeostasis of neuronal Ca^{2+} in age-related cognitive impairment associated with Alzheimer's disease (AD). This is seen when increased oxidative stress and impaired energy metabolism associated with senescent neurons lead to malfunctioning proteins that control membrane excitability and subcellular Ca^{2+} dynamics. Toxic forms of amyloid β-peptide (Aβ) may trigger Ca^{2+} influx into neurons by inducing membrane-associated oxidative stress or forming an oligomeric pore in the membrane, thus, exposing neurons to excitotoxicity and apoptosis. During AD, mutations in the β-amyloid precursor protein and presenilins may compromise the normal proteins in the plasma membrane and endoplasmic reticulum. With time, knowledge of the actions of Ca^{2+} upstream and downstream of Aβ gave a cue to developing some prophylactic or curative interventions for AD [47].

Alzheimer's disease is managed, not cured; the only medications approved for managing the disease are used for mild to moderate AD. These drugs are the cholinesterase inhibitors (ChEI): tacrine, rivastigmine, donepezil, and galantamine. While for moderate to severe AD is memantine, a noncompetitive *N*-methyl-D-aspartate (NMDA) receptor inhibitor, which blocks the excess release of glutamate assumed to be related to cholinergic damage [6, 48, 49].

Research has shown no additional benefit of combination therapy involving vitamin E (2000 IU/day) and selegiline (10 mg/day). However, due to low cost and relative safety, vitamin E was recommended in addition to ChEIs to slow AD progression. Some other treatments, such as Ginkgo biloba, anti-inflammatory drugs, and hormone replacement therapy, have been suggested as possible treatments, although with insufficient evidence [6, 48, 49].

Another treatment option is a combination of memantine and donepezil, and the combination therapy (Namzaric®) was recommended in 2014 to treat individuals with moderate to severe AD who are stabilized on donepezil and memantine therapy. The multi-target-directed ligands (MTDLs) approach currently focuses on designing hybrid molecules that simultaneously regulate multiple biological targets. Moquin is a drug, which has been developed as a potential anti-AD candidate because of its MTDL design capacity. However, combination therapy (CT), including ChEIs and memantine, currently constitutes the best and effective treatment for individuals displaying moderate-to-severe AD. Additionally, CT exhibited better clinical efficacy than monotherapy, along with similar tolerability and safety [6].

3.2 Parkinson disease

Parkinson's disease (PD) is a neurodegenerative clinical syndrome characterized by at least two of four cardinal features: bradykinesia, rigidity, resting tremor, and impairment of postural balance leading to disturbance of gait and falling. James Parkinson, an English physician, was the first to describe this disease, when he called it the "shaking palsy" in 1817 and also coined the term, *paralysis agitans* meaning the shaking palsy, since then, there is still a lack of understanding of the causes of PD [8, 50]. Parkinson's disease may be mistaken with the regular essential tremor; however, the difference in both is that the tremor in Parkinsonism occurs predominantly at rest, while that of essential tremor is seen during actions. Also, tremor in Parkinsonism is unilaterally seen in the arms or legs, while essential tremor bilaterally affects both upper limbs. Bradykinesia is usually the most troublesome symptom. Patients report slowness in performing their daily activities. Falls and swallowing problems are classic signs of late Parkinson's disease; however, if they occur early and are accompanied by unresponsiveness to treatment, they may indicate multiple system atrophy or progressive supranuclear palsy. Early dementia and other features could indicate Lewy body dementia, vascular Parkinsonism, or corticobasal degeneration. Young patients with Parkinsonism (aged <40 years) should always be evaluated for changes in the values of their serum copper and ceruloplasmin levels, with a 24-hour urine collection for copper excretion and slit-lamp examination for Kayser–Fleischer rings in consideration of Wilson's disease [51].

The pathognomonic feature of PD is a loss of the pigmented, dopaminergic neurons of the substantia nigra pars compacta in the brain, with the appearance of intracellular inclusions known as Lewy bodies. During the 1960s, researchers identified a fundamental defect that is a hallmark of the disease: the loss of brain cells that produce an essential chemical, dopamine, which helps direct muscle activity. Gradual loss of dopamine-containing neurons is a feature of normal aging; however, most people do not lose 70–80% of the dopaminergic neurons that cause symptomatic PD. In the absence of treatment, PD gradually deteriorates into a rigid, akinetic state where patients cannot care for themselves within 5–10 years. Death may result from complications of immobility, such as aspiration pneumonia and pulmonary embolism [50, 51].

3.2.1 Diagnosis

This aims to identify ubiquitous Lewy bodies in microscopic postmortem studies, a feature of cell death associated with the disease. However, the clinical diagnosis in PD includes cardinal motor symptoms such as akinesia, rigidity, and tremor [8].

Diagnosis of Parkinsonism involves structured clinical examinations or autopsies. The central pathology in PD is the degeneration of pigmented neurons in the brainstem. Through a microscope, intracellular Lewy bodies are easily identified. The neurons located in the substantia nigra pars compacta are the most affected, resulting in dopamine depletion to its major projection area, the striatum. The depletion culminates into an overactive subthalamic nucleus, which increases the activity of the major inhibitory output nuclei such as the globus pallidus and substantia nigra pars reticulata, resulting in increased inhibition of thalamic activity and problems with motor output. Lewy bodies in the cortex and deeper structures are the main features that distinguish Parkinson's disease (PD) from dementia with Lewy bodies (DLB), a type of neurodegeneration sharing similarities with Alzheimer's and Parkinson's [9].

Due to the presence of Lewy-type α-synucleinopathy in the submandibular glands of PD patients, some scientists considered the feasibility of submandibular gland biopsy for diagnosing PD. Hence, immunohistochemical staining was considered for Lewy-type α-synucleinopathy [10]. Some studies also considered performing needle core biopsies of the submandibular gland in living patients with PD to assess Lewy-type α-synucleinopathy (LTS). Although it was a small-scale study, this tissue biopsy method may be valuable for confirming PD in patients being considered for invasive medical interventions and research studies of other PD biomarkers [52].

Autopsy remains the main definitive diagnostic tool. Some studies provided evidence that unilateral onset of symptoms with features that include tremor and at least one of bradykinesia and rigidity with an efficient initial response to L-dopa have been the best predictors of the pathological diagnosis. In a fifth of the cases, a different neurological disorder was diagnosed at autopsy from that diagnosed during life. Neurological imaging studies with computed tomography or MRI do not reveal any specific changes related to Parkinson's disease. However, most neurologists perform brain imaging tests to rule out rare conditions requiring a different treatment regimen and management strategies, such as normal pressure hydrocephalus or focal lesions. Functional imaging of brain regions affected by Parkinsonism with either positron emission tomography (PET) or single-photon emission tomography SPECT has been proposed [51].

3.2.2 Management

For the efficient management of medical conditions, the risk–benefit ratio is considered; the aim here is to make the patient experience wellness as close to normal function as possible without having side effects from therapy. Hence, the appropriate multi-disciplinary approach must be utilized.

Some factors are considered to determine the optimal choice for the individual patient at different phases. These include the following:

1. Level of patient disability in terms of performing a daily routine which includes work. Here L-dopa may be indicated. Dopamine agonists may be efficient for patients with mild to moderate disabilities. For very mild symptoms, anticholinergic drugs or amantadine may be considered [51].

2. Prevention of response fluctuations. The initial use of dopamine agonists may attenuate the risk of developing dyskinesias, "wearing off" and "on–off fluctuations."

3. Age of the patient. Those that occur at younger ages (aged <65 years) are more tolerable to medications and may bear a lower risk of side effects. Geriatrics often have more challenges with cognitive and psychiatric side effects, and physicians should be cautious while administering anticholinergics and amantadine. Dopamine agonists may also present more side effects in elderly patients [51].

4. Side-effect profile of the drug under review. If a patient is worried about potential drowsiness or may not tolerate a change in mental status or already has cognitive impairment, then a dopamine agonist may not be a good choice. Ankle edema may be exacerbated by amantadine or dopamine agonists [51].

5. Cost for patients without health care coverage. Generic L-dopa/carbidopa and bromocriptine may be the most affordable [51].

Pharmacological attempts to restore dopaminergic activity with levodopa and dopamine agonists have successfully alleviated many of the clinical features of PD. A complementary approach has been to resuscitate the normal balance of cholinergic and dopaminergic influences on the basal ganglia with anticholinergic drugs. The availability of effective pharmacological treatment has drastically altered the prognosis of PD; in many cases, good functional mobility can be achieved for many years, and the life expectancy of well-managed patients is increased substantially [50]. It is important to emphasize that PD therapy must be individualized and tailored to the specific needs of each patient using a basic algorithm [53, 54].

Treatment of early PD with mild symptoms benefit from nonpharmacological therapy such as exercise and relaxation techniques. However, monoamine (MAO) inhibitors such as selegiline, rasagiline, and safinamide; dopamine agonists; or anticholinergic medications are added to ameliorate conditions [50, 53]. Levodopa or dopamine agonist could be added to ease the challenges associated with the motor neurons [50, 53]. Decades of clinical observation have validated levodopa as the most effective primary medicinal agent [50]. Entacapone (Comtan) and rasagiline (Agilect) could help hold brief pending when PD has progressed and the medications seem inefficient in relieving symptoms [50]. Surgical and experimental therapeutics should be considered as the disease progresses and motor complications (including motor fluctuations and dyskinesias) develop [53]. Inosine, which increases urates, and Isradipine, a calcium channel blocker, when added, are treatments designed to prevent the accumulation of toxic α-synuclein. Monoclonal antibodies directed at aggregated α-synuclein in some patients with Parkinson's disease also provided evidence of strong target engagement and CNS penetration [53].

Optimizing the pharmacologic treatment for both motor and non-motor symptoms is critical; however, nutritional interventions cum counseling could also be planned to manage weight gain or loss of weight efficiently. The optimization of levodopa pharmacokinetics and avoidance of interaction with proteins; improvement in gastrointestinal dysfunction such as dysphagia and constipation; prevention and treatment of nutritional deficiencies either the micronutrients or vitamins could systematically be employed [55]. However, other therapeutic interventions such as continuous pump therapies with apomorphine or parenteral levodopa or the implantation of electrodes for deep brain stimulation could also be considered [8]. **Table 3** shows the type of protein detected in some popular neurodegenerative diseases, including those not discussed in this work.

3.3 Huntington disease

The disease got its name from the physician George Huntington, who first described it in late 1872. Huntington's disease is a hereditary, autosomal dominant, progressive neurodegenerative disease associated with a single abnormal gene on chromosome 4. The pathogenesis is initiated by a CAG (glutamine) trinucleotide expansion in exon 1 of the Huntingtin (*HTT*) gene, which is found at the short arm of chromosome 4p16.9. The normal function of the Huntington gene *HTT* is not known, but it may be involved in sustaining the cyclic adenosine monophosphate response element-binding protein, intracellular signaling, and obviating toxicity of neurons. Earlier studies suggest that the conjugation of the striatum-protein-rich Ras homolog with mutant HTT (mHTT) could lead to cellular toxicity. Although, why this protein causes cellular toxicity is poorly understood. Some evidence suggests that the interaction of the mHTT protein and the group 1 metabotropic glutamate receptors may be at the root of the delayed onset [11, 12, 56].

Huntington's disease (HD) is clinically characterized by cognitive dysfunction, abnormal involuntary movements, behavioral disturbance, and psychiatric disease. In abnormal involuntary movements, symptoms may include chorea, dystonia, rigidity, akathisia, bruxism, swallowing disorders, myoclonus, impaired manual dexterity, impaired global motor capacities, and gait and balance disorders. Cognitive dysfunction may present with impaired executive functions, bradyphrenia, language and communication disorders, and social cognition impairments. Behavioral disorders could be associated with memory disorders, disorientation, and visuospatial and visual perceptual disorders. Psychoanalysis of the patient could reveal depression, suicidal ideation or attempts, irritability, apathy, anxiety, obsessions, impulsivity, sexual disorders, hallucinations, sleep disorders, urinary incontinence, pain, dental pain, excessive perspiration, weight loss, hypersalivation, reduced lung function, and respiratory muscle strength [11, 12, 57].

The disease typically lasts 15–20 years, with dementia, mutism, dystonia, and bradykinesia becoming the classic symptoms in advanced forms of the disease. The mean age at onset is between the third and fifth decade of life, with a range of 2–85 years. Juvenile Huntington's disease (JHD) is when the first symptoms and signs appear before the second decade of life. The symptoms of young and old patients vary, as the younger patient presents with an overwhelming rigidity (Westphal variant), while the geriatric becomes bed-bound with rigidity and flexion contractures in the limbs [11, 58].

Pathologically, diffuse neurodegeneration is seen in the cortex and the striatum. The medium spiny neurons are the primary neurons affected, marked with the conspicuous presence of γ-aminobutyric acid and enkephalin. These neurons typically project into the lateral globus pallidus. With time, this degenerative process progresses to the rest of the basal ganglia with subsequent dissemination, reaching the cortex and substantia nigra. Aggregates of mHTT are seen within the nucleus and cytoplasm during microscopy. Inclusion bodies containing a complex of Huntingtin and other soluble mHTT are seen in the neurons. The cause of the cell death seen in this disease is yet to be delineated between the accumulation of the mHTT conglomerate or the soluble form of the protein when toxic. Glutamate, dopamine, and γ-aminobutyric acid are considered the most affected neurotransmitters in HD; hence, they are the focus of current pharmacological interventions [11, 12, 58].

The prevalence among the European is at 4–8 cases per 100,000, While America has not had a general epidemiological study since 1993. It is rare in Japan and Finland but common in Scotland and Venezuela. At the same time, there are inadequate data from Africans, Black Americans, and those in Eastern Asia [58].

3.3.1 Diagnosis

After accessing the clinical signs and symptoms presented, a genetic test called predictive test is requested. A DNA test showing abnormal CAG expansion (or repeats) in the *HTT* gene can be used to confirm the diagnosis in symptomatic individuals. The CAG (cytosine (C), adenine (A), and guanine (G)) repeats seen in the juvenile HD is over 55 in most cases. While for the elderly, it is about 36–40. The longer the repeats, the younger the age of the patients [11]. Biomarkers could also be exploited. With biomarkers, identification of mHTT in CSF could be a positive indication [12].

3.3.2 Management

Due to the myriad of disorders involved in this disease, it benefits more from symptomatic management rather than a definitive cure. This symptomatic management which is multi-disciplinary in approach, includes physical therapy, gastrostomy device, and medications such as antidepressants and antipsychotics—the chorea benefits from atypical antipsychotic drugs, which include olanzapine and tetrabenazine. Irritability benefits from an atypical antipsychotic drug in severe cases, but in mild cases, the use of selective serotonin reuptake inhibitor (SSRI), an antidepressant, may suffice. For obsessive–compulsive thoughts and actions, experts recommend SSRIs. For dystonia, physical therapy and injection of botulinum toxin are advocated [11, 58].

Since this disease involves the production of aberrant proteins, targeting the DNA or RNA may form a basis for drug discovery [12].

4. Metabolic disorders

Most metabolic disorders are nutrition-based disorders that are a result of the diet and lifestyle of the patients. Among these nutrition-based diseases, some ailment result from undernutrition due to food scarcity, leading to insufficient energy. At the same time, some result from overnutrition due to the insufficient capacity of the hormones regulating these nutrients. In this section, the focus is made on protein-energy malnutrition and diabetes mellitus, a type of hormonopathy.

4.1 Protein-energy malnutrition

The World Health Organization considers malnutrition in the context of both undernutrition and overnutrition. It could be described as the cellular imbalance between the supply of nutrients and energy and the body’s demand to ensure healthy development, maintenance, and specific functions [59]. The term protein-energy malnutrition (PEM) includes kwashiorkor, marasmus, and intermediate states of marasmic-kwashiorkor. Those below 5 years may present a mixed picture of marasmus and kwashiorkor or milder forms of malnutrition. Malnutrition among children leads to waned immunity and increases susceptibility to diseases. Inadequate access to nutritious foods due to rising food prices is a common cause of malnutrition [60]. In the former times, rising food prices or food scarcity was induced by war; however, in recent times, terrorism and climate change could be implicated. Areas close to the desert may be experiencing an acute food shortage due to severe drought and other effects of climate change.

Studies have shown that those between 6 and 12 months are most affected, with over half of this population studied presenting with PEM and a third of those 13–24 months having PEM. Among these studies, marasmus is the most prevalent form of PEM, affecting a third of the population studied. Diarrhea and malaria are the associated co-morbidities popular with this disease, with over 60% of these populations coming from the lower socioeconomic status. The case fatality rate was 40.1%, with the males having more prevalence at 50.9%. Mortality among the marasmic-kwashiorkor and the unclassified group was 53.3 and 54.5%, respectively [61, 62].

The World Health Organization estimates that about two-thirds of all deaths occurring among pediatrics in developing countries could be attributed to malnutrition. Therefore, improving nutrition is crucial for reducing high infant and under-five mortality rates, the proportionate physical growth, the social and mental well-being of children, and academic achievement [61, 62]. Sub-Saharan Africa suffers the most from PEM around the world.

4.1.1 Diagnosis

The assay of total protein and albumin helps diagnose PEM, as early detection helps obviate the challenges associated with severe forms of PEM. In the final stage of wasting, reduced plasma albumin concentration ensues due to the adaptation of the human system to a protein-deficient diet. The development of marasmus reveals energy deficiencies in the diet, which leads to the change of the regular pattern of proteins. It is also observed that a decrease in serum albumin and total protein in PEM was due to reduced synthesis of protein resulting from inadequate intake of dietary protein. PEM in children is associated with a more significant deficiency of total protein, which may be as low as 50% of the child's total protein in severe cases. These reductions of total serum protein and albumin are prominent in kwashiorkor and marasmus [63].

In the absence of a diagnostic facility, the nutritional status of children is determined by clinical examination, history, and anthropometric measurements, which include height-for-age, weight-for-age, weight-for-height, head circumference, mid-upper arm circumference, and skinfold thickness which could be compared to the reference charts of the World Health Organization [60].

4.1.2 Management

A healthy balanced meal is advocated to fortify the immunity of infants and children under the age of 2. However, due to the inability to decipher the best description of a healthy balanced meal for children of such ages in most rural areas, exclusive breastfeeding is strongly advised for the first 6 months after delivery. While locally available meals and fruits rich in vitamin C, coupled with proteins such as eggs, are augmented with breastfeeding between the first 6 and 24 months of delivery. This daily intake of an egg and vitamin C-rich foods (or tablets) for at least one month is based on the need to boost the immunity and replenish the worn-out tissues of these pediatrics [64, 65]. The prevention of PEM cannot be overemphasized as it is associated with a high mortality rate among children under the ages of five in Sub Sahara, Africa [61, 62].

4.2 Hormonopathy

Hormonopathy is a term used in describing a disease associated with a change in the production of hormones. In these conditions, there can be over-secretion or under-secretion of hormones or even the production of aberrant proteins. Endocrine proteinopathies, which are grouped under hormonopathy, are diseases associated with peptide- or protein-derived hormones. They are characterized by the hyper- or hypo-secretion of these proteins or an aberration in their structure and function. Protein-derived hormones include insulin, prolactin, ACTH, gastrin, parathyroid hormone, oxytocin, leptin, ADH and growth hormone. This section focuses on the most prevalent endocrine proteinopathy related to insulin, a disease called diabetes mellitus.

4.2.1 Diabetes mellitus

Diabetes causes severe life-threatening complications, such as hyperglycemic coma, hypoglycemic coma, severe impairment of renal function, blurred vision, memory loss, insulin allergy, and acute neuropathy. Managing it requires dietary control, physical exercise, and insulin administration. Demographic data is based on the patient's age, sex, location, and income. Clinical data is divided into physical signs and laboratory results. Physical signs are those obtained via physical examination of the patient, like BMI (body-mass index), pulse rate, and blood pressure, while the laboratory results are based on the blood sugar levels [66, 67].

An expert system determined by a set of rules used to make decisions is known as a rule-based expert system. Developing this expert system requires a knowledge of the engineering process in which the rules used by human experts are collated and translated into an appropriate form for computer processing [66, 67]. The rule-based expert system is utilized in this section.

4.2.2 Diagnosis

Laboratory results are associated with laboratory tests, like blood and urine tests [66–70]. Using the rule-based expert system, diagnosis is classified into:

I. Test urine for glucose and ketones.

II. Measure random or fasting blood glucose:

- Fasting plasma glucose ≥7.0 mmol/L
- Random plasma glucose ≥11.0 mmol/L

III. Oral glucose tolerance test:

- Fasting plasma glucose 6.1–6.9 mmol/L
- Random plasma glucose 7.0–11.0 mmol/L

4.2.3 Management

The management of diabetes can be classified into three categories.

I. Type-I (juvenile) Diabetes- only Insulin is used.

II. Gestational Diabetes- only Insulin is used, and

DOI: http://dx.doi.org/10.5772/intechopen.101051

III. Type-II Diabetes- the oral hypoglycemic agent, low-carbohydrate diet, and sometimes Insulin is used. In this category, there are unique treatments:

- The drugs for obese and lean patients.
- The specific drugs for patients with challenges with their organs, such as renal diseases, lactic acidosis, and liver disease [69, 70].

5. Conclusion

Early and accurate diagnosis of protein-related diseases saves cost and prevents rapid deterioration of these disease conditions. It also prevents the waste of time and resources used in managing a misdiagnosed condition. The exposition in this chapter should be an eye-opener to the public and stakeholders in the public health domain to harp on the need for early diagnosis, treatment, or management of these diseases for improved health for all.

Health professionals and researchers are encouraged to give further research attention towards discovering a cure and treatment of these protein-related diseases. This can be done by utilizing the knowledge garnered from protein synthesis and the post-translational modification of proteins.

This work which is a summary of the prevalent neurodegenerative diseases and some metabolic disorders has taken the first step to elucidate on how proteins formed the basis of some modalities involved in the diagnosis and treatment of a few of these diseases; the onus lies on researchers in this field to consolidate on it and bring succor to the ailing population.

Acknowledgements

Special thanks to all those who diagnose and manage protein-related diseases in Nigeria; to those children of Northern Nigeria who became the first experience in PEM; to the geriatrics in Mgbidi, Oru west, whose health challenges continue to be an inspiration.

Conflict of interest

The authors declare no competing interests.

Notes/thanks/other declarations

Ohanube A.K. Goodluck conceptualized the work and wrote the section on neurodegenerative diseases and protein-energy malnutrition; Obeta M. Uchejeso wrote the section on Hormonopathy and Introduction; Ikeagwulonu Richard wrote the abstract and conclusion. All authors reviewed the work and approved the final manuscript.

Author details

Ohanube A.K. Goodluck[1], Obeta M. Uchejeso[2*] and Ikeagwulonu R. Chinaza[3]

1 Department of Biomedical Sciences, The Hochschule Bonn-Rhein-Sieg, University of Applied Sciences, Bonn, Germany

2 Department of Medical Laboratory Management, Federal School of Medical Laboratory Science, Jos, Nigeria

3 Department of Medical Laboratory Services, Alex Ekwueme Federal University Teaching Hospital, Abakaliki, Nigeria

*Address all correspondence to: uchejesoobeta@gmail.com

References

[1] Amor S, Puentes F, Baker D, Van Der Valk P. Inflammation in neurodegenerative diseases. Immunology. 2010;**129**(2): 154-169

[2] Prusiner SB. Neurodegenerative diseases and prions. New England Journal of Medicine. 2001;**344**(20): 1516-1526

[3] Sánchez-González L, Maddox RA, Lewis LC, Blevins JE, Harker EJ, Appleby BS, et al. Human prion disease surveillance in Washington State, 2006-2017. JAMA Network Open. 2020;**3**(10):e2020690

[4] Maddox RA, Person MK, Blevins JE, Abrams JY, Appleby BS, Schonberger LB, et al. Prion disease incidence in the United States: 2003-2015. Neurology. 2020;**94**(2):e153-e157

[5] Centre for Disease Control. Creutzfeldt–Jacob Disease. Available from: https://www.cdc.gov/prions/cjd/occurrence-transmission.html [Accessed: 11 August 2021]

[6] Kabir M, Uddin M, Mamun AA, Jeantet P, Aleya L, Mansouri RA, et al. Combination drug therapy for the management of Alzheimer's disease. International Journal of Molecular Sciences. 2020;**21**(9):3272

[7] Doecke JD, Laws SM, Faux NG, Wilson W, Burnham SC, Lam CP, et al. Blood-based protein biomarkers for diagnosis of Alzheimer disease. Archives of Neurology. 2012;**69**(10):1318-1325

[8] Pedrosa DJ, Timmermann L. Review: Management of Parkinson's disease. Neuropsychiatric Disease and Treatment. 2013;**9**:321-340. DOI: 10.2147/NDT.S32302

[9] Frank C, Pari G, Rossiter JP. Approach to diagnosis of Parkinson's disease. Canadian Family Physician. 2006;**52**(7): 862-868

[10] Beach TG, Adler CH, Dugger BN, Serrano G, Hidalgo J, Henry-Watson J, et al. Arizona Parkinson's disease consortium. Submandibular gland biopsy for the diagnosis of Parkinson disease. Journal of Neuropathology and Experimental Neurology. 2013;**72**(2): 130-136

[11] Roos RA. Huntington's disease: A clinical review. Orphanet Journal of Rare Diseases. 2010;**5**(1):1-8

[12] McColgan P, Tabrizi SJ. Huntington's disease: A clinical review. European Journal of Neurology. 2018;**25**(1):24-34

[13] Wadsworth JD, Collinge J. Update on human prion disease. Biochimica et Biophysica Acta (BBA) - Molecular Basis of Disease. 2007;**1772**(6):598-609

[14] Safar JG, Geschwind MD, Deering C, Didorenko S, Sattavat M, Sanchez H, et al. Diagnosis of human prion disease. Proceedings of the National Academy of Sciences. 2005;**102**(9):3501-3506

[15] Bellon A, Seyfert-Brandt W, Lang W, Baron H, Gröner A, Vey M. Improved conformation-dependent immunoassay: Suitability for human prion detection with enhanced sensitivity. Journal of General Virology. 2003;**84**(7):1921-1925

[16] Graziano S, Pocchiari M. Management and prevention of human prion diseases. Current Neurology and Neuroscience Reports. 2009;**9**(6):423

[17] Wille H, Requena JR. The structure of PrP^{Sc} prions. Pathogens. 2018;7(1):20. DOI: 10.3390/pathogens7010020

[18] Hughes D, Halliday M. What is our current understanding of PrP^{Sc}-associated neurotoxicity and its molecular underpinnings? Pathogens. 2017;**6**(4):63. DOI: 10.3390/pathogens6040063

[19] Brownell B, Oppenheimer DR. An ataxic form of subacute presenile polioencephalopathy (Creutzfeldt–Jakob disease). Journal of Neurology, Neurosurgery, and Psychiatry. 1965;**28**(4):350

[20] Matsue E, Kinoshita T, Sugihara S, Fujii S, Ogawa T, Ohama E. White matter lesions in panencephalopathic type of Creutzfeldt–Jakob disease: MR imaging and pathologic correlations. American Journal of Neuroradiology. 2004;**25**(6): 910-918

[21] Gambetti P, Kong Q, Zou W, Parchi P, Chen SG. Sporadic and familial CJD: Classification and characterization. British Medical Bulletin. 2003;**66**(1): 213-239

[22] Collinge J. Human prion diseases and bovine spongiform encephalopathy (BSE). Human Molecular Genetics. 1997;**6**(10):1699-1705

[23] Collinge J. Prion diseases of humans and animals: Their causes and molecular basis. Annual Review of Neuroscience. 2001;**24**(1):519-550

[24] Riek R, Wider G, Billeter M, Hornemann S, Glockshuber R, Wüthrich K. Prion protein NMR structure and familial human spongiform encephalopathies. Proceedings of the National Academy of Sciences. 1998;**95**(20):11667-11672

[25] Swietnicki W, Petersen RB, Gambetti P, Surewicz WK. Familial mutations and the thermodynamic stability of the recombinant human prion protein. Journal of Biological Chemistry. 1998;**273**(47):31048-31052

[26] Montagna P, Gambetti P, Cortelli P, Lugaresi E. Familial and sporadic fatal insomnia. The Lancet Neurology. 2003;**2**(3):167-176

[27] Collinge J, Sidle KC, Meads J, Ironside J, Hill AF. Molecular analysis of prion strain variation and the etiology of new variant CJD. Nature. 1996;**383**(6602):685-690

[28] Panegyric PK, Armari E. Therapies for human prion diseases. American Journal of Neurodegenerative Disease. 2013;**2**(3):176-186

[29] Collinge J, Whitfield J, McKintosh E, Beck J, Mead S, Thomas DJ, et al. Kuru in the 21st century—An acquired human prion disease with very long incubation periods. Lancet. 2006;**367**:2068-2074

[30] Mead S, Stumpf MP, Whitfield J, Beck JA, Poulter M, Campbell T, et al. Balancing selection at the prion protein gene consistent with prehistoric kuru-like epidemics. Science. 2003;**300**(5619): 640-643

[31] Lee HS, Brown P, Cervenáková L, Garruto RM, Alpers MP, Gajdusek DC, et al. Increased susceptibility to Kuru of carriers of the PRNP 129 methionine/methionine genotype. The Journal of Infectious Diseases. 2001;**183**(2):192-196

[32] Will RG, Ironside JW, Zeidler M, Estibeiro K, Cousens SN, Smith PG, et al. A new variant of Creutzfeldt–Jakob disease in the UK. The Lancet. 1996;**347**(9006):921-925

[33] Sivitz L, Erdtmann R, editors. Advancing Prion Science: Guidance for the National Prion Research Program: Interim Report. Washington DC: National Academies Press; 2003

[34] Wang J, Wang X, Gao X, Vortmeyer AO. Prion diseases and their PrPSc-based molecular diagnostics. Journal of Neurology and Neuroscience. 2002;**4**(6):56

[35] Riesner D. Biochemistry and structure of PrPC and PrPSc. British Medical Bulletin. 2003;**66**(1):21-33

[36] Chen C, Dong X. Therapeutic implications of prion diseases. Biosafety and Health. 2021;**3**(2)

[37] Skinner PJ, Seelig DM. Past, Present, and Potential Future Prion Disease Treatment Strategies. Prion: An Overview. UK: IntechOpen; 2017. p. 27

[38] Elezgarai SR, Biasini E. Common therapeutic strategies for prion and Alzheimer's diseases. Biological Chemistry. 2016;**397**(11):1115-1124

[39] Abdulrahman BA, Tahir W, Doh-Ura K, Gilch S, Schatzl HM. Combining autophagy stimulators and cellulose ethers for therapy against prion disease. Prion. 2019;**13**(1):185-196

[40] Petrella JR, Coleman RE, Doraiswamy PM. Neuroimaging and early diagnosis of Alzheimer disease: A look to the future. Radiology. 2003;**226**(2):315-336

[41] Sanabria-Castro A, Alvarado-Echeverría I, Monge-Bonilla C. Molecular pathogenesis of Alzheimer's disease: An update. Annals of Neurosciences. 2017;**24**(1):46-54

[42] Johnson DK, Wilkins CH, Morris JC. Accelerated weight loss may precede diagnosis in Alzheimer's disease. Archives of Neurology. 2006;**63**(9): 1312-1317

[43] Zhao Y, Zhao B. Oxidative stress and the pathogenesis of Alzheimer's disease. Oxidative Medicine and Cellular Longevity. 2013:1-8

[44] Fan L, Mao C, Hu X, Zhang S, Yang Z, Hu Z, et al. New insights into the pathogenesis of Alzheimer's disease. Frontiers in Neurology. 2020;**10**:1312

[45] Beach TG, Monsell SE, Phillips LE, Kukull W. Accuracy of the clinical diagnosis of Alzheimer disease at National Institute on Aging Alzheimer Disease Centers, 2005-2010. Journal of Neuropathology and Experimental Neurology. 2012;**71**(4):266-273

[46] Waldemar G, Dubois B, Emre M, Georges J, McKeith IG, Rossor M, et al. Recommendations for the diagnosis and management of Alzheimer's disease and other disorders associated with dementia: EFNS guideline. European Journal of Neurology. 2007;**14**(1):e1-e26

[47] Bezprozvanny I, Mattson MP. Neuronal calcium mishandling and the pathogenesis of Alzheimer's disease. Trends in Neurosciences. 2008;**31**(9): 454-463

[48] Zhu CW, Sano M. Economic considerations in the management of Alzheimer's disease. Clinical Interventions in Aging. 2006;**1**(2): 143-154. DOI: 10.2147/ciia.2006.1.2.143

[49] Stokin GB, Lillo C, Falzone TL, Brusch RG, Rockenstein E, Mount SL, et al. Axonopathy and transport deficits early in the pathogenesis of Alzheimer's disease. Science. 2005;**307**(5713): 1282-1288

[50] Goldenberg MM. Medical management of Parkinson's disease. Pharmacy and Therapeutics. 2008; **33**(10):590

[51] Guttman M, Kish SJ, Furukawa Y. Current concepts in the diagnosis and

management of Parkinson's disease. CMAJ. 2003;**168**(3):293-301

[52] Adler CH, Dugger BN, Hinni ML, Lott DG, Driver-Dunckley E, Hidalgo J, et al. Submandibular gland needle biopsy for the diagnosis of Parkinson disease. Neurology. 2014;**82**(10):858-864

[53] Tarakan A, Jankovic J. Diagnosis and management of Parkinson's disease. Seminars in Neurology. 2017;**37**(2): 118-126

[54] Nutt JG, Wooten GF. Diagnosis and initial management of Parkinson's disease. New England Journal of Medicine. 2005;**353**(10):1021-1027

[55] Barichella M, Cereda E, Pezzoli G. Major nutritional issues in the management of Parkinson's disease. Movement Disorders. 2009;**24**(13): 1881-1892

[56] Duff K, Paulsen JS, Beglinger LJ, Langbehn DR, Wang C, Stout JC, et al. "Frontal" behaviors before the diagnosis of Huntington's disease and their relationship to markers of disease progression: Evidence of early lack of awareness. The Journal of Neuropsychiatry and Clinical Neurosciences. 2010;**22**(2):196-207

[57] Johnson SA, Stout JC, Solomon AC, Langbehn DR, Aylward EH, Cruce CB, et al. Beyond disgust: Impaired recognition of negative emotions prior to diagnosis in Huntington's disease. Brain. 2007;**130**(7):1732-1744

[58] Bachoud-Lévi AC, Ferreira J, Massart R, Youssov K, Rosser A, Busse M, et al. International guidelines for the treatment of Huntington's disease. Frontiers in Neurology. 2019;**10**:710

[59] World Health Organization Nutrition for Health and Development Protection of the Human Environment Geneva 2005. Malnutrition: Quantifying the health impact at national and local levels. Available from: https://www.who.int/quantifying_ehimpacts/publications/MalnutritionEBD12.pdf [Accessed: 6 August 2021]

[60] Akugizibwe R, Kasolo J, Makubuya DB, Damani AM. Missed opportunities in the diagnosis and management of protein-energy malnutrition among children under 5 years in Wakiso district, Uganda. Journal of Public Health and Epidemiology. 2013;**5**(11):463

[61] Ubesie AC, Ibeziako NS, Ndiokwelu CI, Uzoka CM, Nwafor CA. Under-five protein-energy malnutrition admitted at the University of in Nigeria teaching hospital, Enugu: A 10 year retrospective review. Nutrition Journal. 2012;**11**(1):1-7

[62] Hamer C, Kvatum K, Jeffries D, Allen S. Detection of severe protein-energy malnutrition by nurses in The Gambia. Archives of Disease in Childhood. 2004;**89**(2):181-184

[63] Chowdhury MS, Akhter N, Haque M, Aziz R, Nahar N. Serum total protein and albumin levels in different grades of protein-energy malnutrition. Journal of Bangladesh Society of Physiologist. 2008;**3**:58-60

[64] Ohanube GA, Obeta MU, Ikeagwulonu RC, Jwanse IR. COVID-19: A case study of using vitamin C enriched plants and ascorbic acid as cure. American Journal of Case Reports. 2020;**8**(11):435-437

[65] Ohanube GA, Obeta UM, Ikeagwulonu CR. Case reports in the use of vitamin C based regimen in prophylaxis and management of COVID-19 among Nigerians. Journal of Current Biomedical Reports. 2020;**1**(2):77-80

[66] Akter M, Uddin MS, Haque A. Diagnosis and management of diabetes mellitus through a knowledge-based system. In: 13th International Conference on Biomedical Engineering. Berlin, Heidelberg: Springer; 2009. pp. 1000-1003

[67] Mpondo BC, Ernest A, Dee HE. Gestational diabetes mellitus: Challenges in diagnosis and management. Journal of Diabetes and Metabolic Disorders. 2015;**14**(1):1-7

[68] Chamberlain JJ, Rhinehart AS, Shaefer CF Jr, Neuman A. Diagnosis and management of diabetes: Synopsis of the 2016 American Diabetes Association Standards of Medical Care in Diabetes. Annals of Internal Medicine. 2016;**164**(8): 542-552

[69] Baynes HW. Classification, pathophysiology, diagnosis, and management of diabetes mellitus. Journal of Diabetes & Metabolism. 2015;**6**(5):1-9

[70] Sacks DB, Bruns DE, Goldstein DE, Maclaren NK, McDonald JM, Parrott M. Guidelines and recommendations for laboratory analysis in the diagnosis and management of diabetes mellitus. Clinical Chemistry. 2002;**48**(3):436-472

Chapter 3

Methods of Protein Detection in Cancer for Diagnosis, Prognosis and Therapy

Shenbagamoorthy Sundarraj, Gopalan Rajagopal, Balaji Sundaramahalingam, Madasamy Sundar and Ramar Thangam

Abstract

Emerging proteomic technologies offer new insight in the study of malignant tumor to identify protein biomarkers for early detection, stratification, prediction and monitoring of treatment, as well as to detect target molecules for therapy. The tumor protein biomarker is responsible for the regulation of the cell cycle to promote cell proliferation and resistance to cell death. Important technologies include ELISA, immunohistochemistry, flow cytometry, western blot, mass spectrometry, protein microarray, and microfluidics for the study of screening, protein profiling, identification, qualitative and quantitative analysis of differential expressed oncoproteins relative to cancer tissues, counterparts at different stages of the disease from preneoplasia to neoplasia. It can also provide a detailed description of identifying tissue-specific protein biomarkers and to analysis the modification of protein activity in cancer conditions. In this chapter, we discuss current and emerging protein assays for improving cancer diagnosis, including trends toward advances in assay miniaturization, improve sensitivity and specificity, time and cost-effective, and accuracy in detection and measurement of protein activity. However, information from these protein diagnostic technologies should be integrated to obtain the optimal information required for the clinical management of a patient.

Keywords: cancer, protein biomarker, ELISA, protein microarray, mass spectrometry

1. Introduction

Cancer is the leading cause of death global population. As stated by the National Cancer Institute annual report revealed that there were 18.1 million new cases and 9.5 million cancer-related deaths globally in 2018. By 2040, the number of new cancer cases per year is anticipated to arise around 29.5 million and the number of cancer-related deaths to 16.4 million [1]. Cancer mortality can be reduced if cases are detected earlier and treated systematically and can result in a greater probability of survival rate and less morbidity [2]. Cancer diagnosis and prognosis have advanced dramatically during the last decades. Achieving this goal will necessitate not only improved therapies, but also enhanced methods for evaluating an individual's risk of developing cancer, detecting cancers at an early stage when they can be treated more effectively, distinguishing aggressive from non-aggressive cancers, and monitoring recurrence and response to therapy.

Diagnostic imaging technologies can be used to detect people with cancer, these tests can be physically invasive, time-consuming and expensive to screen large groups of people who are asymptomatic and can cause unnecessary stress and worry. Furthermore, diagnostic imaging technologies frequently overlook minor lesions, resulting in the disease not being detected until it has progressed to the point when treatment intervention is less effective. However, insufficient diagnostics prohibit the detection of certain types of cancer until the advanced stage. For instance endoscopy with biopsy is the distinctive screening method for esophageal cancer and is generally performed after symptoms appear [3]. Other screens may be providing high levels of false positives or negatives. Hepatocellular carcinoma is generally detected by ultrasound, but this technique is subjected to operator mistake and often cannot distinguish between malignant and benign nodules [4]. Although mammography is the standard screening techniques for breast cancer, 20% of cases go undetected with this screening and specificity is 25%, leading to a large number of false positives [5].

Improving strategies for screening asymptomatic individuals for early-stage malignancies is a particularly difficult challenge. Overcome these challenges in the recent years, there has been a surge in interest in molecular markers as a cancer diagnosis, prognosis and therapeutic response [6]. The cancer antigen 15–3 (CA 15–3) act as a potential biomarker is used to screening and monitoring breast cancer [7]. The prostate-specific antigen (PSA) is a widely mentioned marker that is used to test male patients for prostate cancer [8]. The analysis of overexpression of human epidermal growth factor receptor type 2 (Her2) and estrogen receptor levels in breast cancer patients [9, 10].

Specifically, some modern molecule-oriented techniques used protein as a biomarker for monitoring of cancer progression and early tumor detection. Furthermore, tumor biomarker protein assays are suitable method for holding important clinical diagnostic tests in future because gene level studies may not correlation for the cancer alteration [11, 12]. Protein biomarkers played significant roles in accurate early diagnosis, therapy and prognosis in colorectal cancer [13].

Serum protein biomarkers are well developed tools for cancer diagnosis [14]. As proof, prostate-specific antigen (PSA), cancer antigen 125 (CA-125) and carcinoembryonic antigen (CEA) is extensively used for the diagnosis and management of various types of cancer, namely prostate, ovarian and gastrointestinal cancers [15]. The clinical sensitivity of a biomarker can simply be defined as the proportion of people with a confirmed disease who test positive for the biomarker assay whereas specificity refers to the proportion of healthy individuals who test negative for the biomarker assay [16]. Noninvasive assays, such as those using blood, stool, urine, or saliva are preferred because they cause less pain to the patient, have higher abidance rates and may be taken frequently for monitoring the treatment response [17]. Measurement of serum proteins ensures distinguishes between various types of malignancies from benign and thus leading imaging analysis, endoscopic examination and other diagnostic procedures and monitoring of the efficacy of the treatment.

The main objective of this chapter is to provide a new insight of emerging technologies based on protein detection for cancer diagnostic and prognostic. Proteins analytical techniques are especially suitable for the diagnosis of cancer are described here briefly along with their recent development. We also provide a brief description of techniques currently used to identify the protein activity in post-translational modification have been linked to cancer diagnosis and cancer progression.

2. Methods of proteins detection in cancer

The developments of proteomic patterns have emerged as effective method for diagnostics in the field of cancer proteomics in that it is represents new way for cancer detection and also clinically feasible. The enzyme-linked immunosorbent assay (ELISA), immunohistochemistry (IHC) and flow cytometry system represent the most reliable, sensitive and widely available protein-based testing platform in the clinic for the diagnosis, prognosis and treatment monitoring of cancer [18, 19]. Other protein analyses techniques such as mass spectrometry, Protein array and Microfluidics are currently at the laboratory level setup extensively used for cancer research purpose but these techniques are being developed for clinical applications (**Figure 1** and **Table 1**).

2.1 Enzyme-linked immunosorbent assay

The enzyme-linked immunosorbent test (ELISA) has been widely utilised in regular clinical diagnostics and is still considered the gold standard for detecting

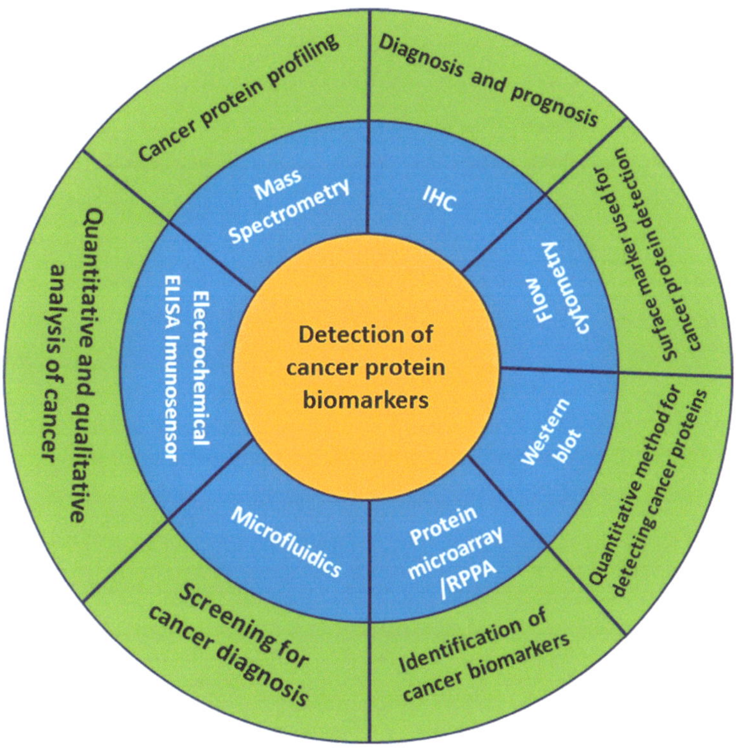

Figure 1.
Schematic illustration of the different proteomic techniques for detection, identification, screening, protein profiling and modification of protein in various cancers.

Cancer Protein Biomarker	Cancer type	Specimen	Methodology	Reference
Midkine	Hepatocellular carcinoma	Serum	ELISA	[20]
Alpha-fetoprotein (AFP)	Liver cancer	Serum	Electrochemical immunosensor, ELISA	[21]
Midkine	Cancer cells	Serum	Electromagnetically induced transparency (EIT)	[22]
EGFR, CA125 and HE4	Ovarian cancer	Serum	CIMA, ELISA	[23]
Vim3, Mxi-2	Renal carcinoma	Urine	ELISA	[24]
Cyclin D1	Breast cancer	Tissue	WB, IHC	[25]
BAP31	Ovarian cancer	Tissue	IHC	[26]
MDH2	NSCLC	Urine	ELISA	[27]
Aminoacylase-1 (ACY-1)	Rectal cancer	Tissue	IHC	[28]
MMP-1	Oral cancer	Saliva, Urine	Electrochemical immunosensor	[29]
Tumor-associated antigens (TAAbs)	Lung cancer	Serum	Protein array, ELISA	[30]
HPV E6 and E7 oncoproteins	Cervical cancer	Serum	Electrospun PCL (ePCL) Fiber coated ELISA	[31]
RAS Q61R	Epithelial-Myoepithelial Carcinoma	Tissue	IHC	[32]
Carcinoembryonic antigen, CA 15–3	Breast cancer	Serum	Electrochemical aptasensor Redox probes labeled aptamers	[33]
Integrin alpha V	Prostate cancer	Urine	ELISA	[34]
B-cell activating factor (BAFF)	Melanoma	Serum	ELISA, IHC	[35]
QSOX1	Colorectal Cancer	Serum	ELISA	[36]
LRG1, TTR, CA 19–9	Pancreatic cancer	Plasma	ELISA	[37]
Carcinoembryonic antigen		Serum	MALDI-TOF MS	[38]
Tumor-associated antigens (TAAbs)	Gastric Cancer	Serum	Proteomic chips, ELISA	[39]
PD-L1, HIF-1α	Breast cancer	Serum	Electrochemical immunosensor	[40]
FGL1		Serum	Nanobody-based ELISA	[41]
p16INK4a	Cervical cancer	cervical swabs	ELISA, IHC, WB	[42]

Cancer Protein Biomarker	Cancer type	Specimen	Methodology	Reference
FBLN1, ANT3	Cervical cancer	Serum, Plasma, tissue	ELISA, LC MS/MS	[43]
ATX and LPA	Pancreatic cancer	Serum	ELISA	[44]
Tumor Endothelial	Colorectal cancer	Tissue, Serum	IHC, ELISA	[45]
Marker 8 (ANTXR1) SAS1B	Thyroid cancer	Serum	ELISA	[46]
Annexin A2	Ovarian Cancer	Plasma	ELISA	[47]
DKK3	Ovarian Cancer	Serum	ELISA	[48]
Alpha-fetoprotein	Liver cancer	Human cord serum	ELISA	[29]
Desmoglein 3 (DSG3)	Head and neck squamous cell carcinoma	Cell lysate	Microfluidic immunoarray	[49]
FGFR3	Bladder cancer	Recombinant FGFR3 protein	Electrochemical impedance spectroscopy, ELISA	[50]
BC2L-C lectin	Breast cancer	Tissue	IHC	[51]
MUCIN-16/CA125	Ovarian Cancer	Plasma	MS	[52]
Matrix metalloproteinase-1	Oral cancer	Saliva	MS	[53]
HER2 and Ki67	Breast cancer	Tissue	Western blot	[9]
Perilipin-2	Renal cancer	Urine	ELISA, Bioplasmonic paper–based assay	[54]
apo A-IV and LRG1	Oral cancer	Plasma	ELISA, WB, LC MS/MS	[55]
NANOG	Ovarian cancer	Tissue	IHC, WB	[56]
MUCIN-16/CA125	Ovarian Cancer	Plasma	Proximity extension assay	[15]
ER	Breast cancer	Tissue	IHC	[10]
rab31 and mucin-1 (CA15–3)	Breast cancer	Tissue	ELISA, chemiluminescence immunoassay	[57]
HER2	Breast cancer	Tissue	IHC-QD	[58]
PSA and IL-6	Prostate cancer	Serum	Microfluidic electrochemical immunoassay	[59]

Table 1.
List of cancer protein biomarkers and their detection methods in the current research practice.

cancer protein biomarkers in physiological samples [60–62]. Despite advances in developing cost-effective and label-free novel ELISA-based methods for future use in point-of-care cancer diagnostics, prognostics and therapy monitoring offers promise for improving the early detection of breast cancer (**Figure 2**). In conventional ELISA

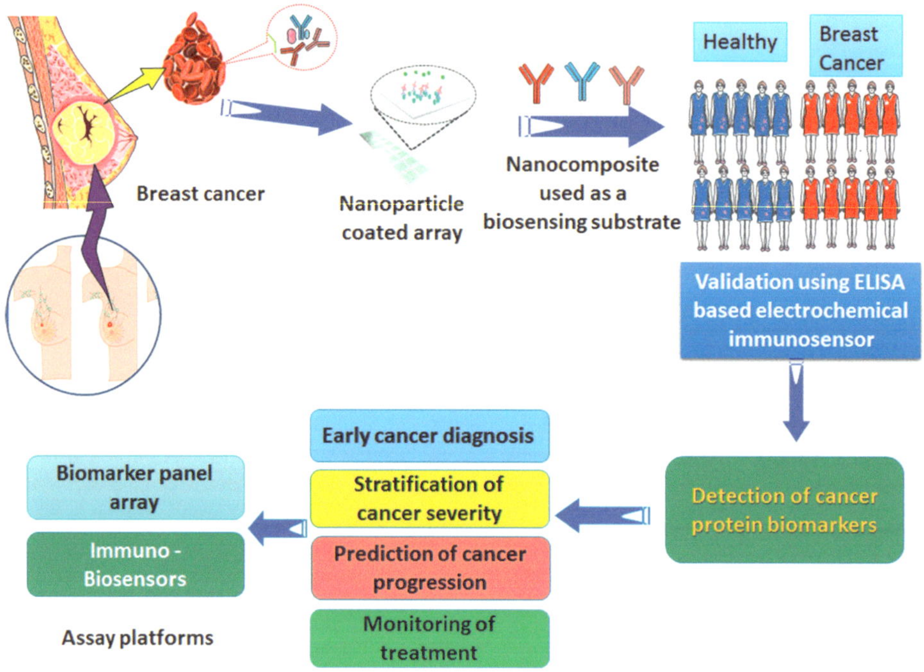

Figure 2.
Flowchart of nanocomposite used as a biosensing substrate and detection in breast cancer using nanoparticle coated arrays and ELISA based electrochemical immunosensor. First, a drop of blood from breast cancer (BC) patients is subjected to an nanoparticle fabricated array for a high-throughput screening of cancer biomarkers in breast cancer. Second, promising cancer biomarker candidates are selected from the array screening and validated in a large cohort of patients using ELISA, which can be used for early diagnosis, disease stratification, prediction of disease progression, or monitoring of drug responses. Finally, according to the function of nanoparticle coated array biomarker panel, biosensors could be designed and fabricated for clinical use in breast cancer.

techniques, colorimetric or fluorescent readout signals are utilised to show FDA-approved protein biomarkers currently used in clinical practice (**Table 2**).

Stevens et al. generated a novel ELISA-based technique called plasmonic ELISA that uses gold nanoparticles as the probe to detect PSA in prostate cancer diagnosis [63, 64]. Shim et al. described a microfluidic droplet-based extremely flexible and sensitive diagnostic device for counting individual analyse molecules and identifying a biomarker for prostate cancer in buffer with a detection limit of 46 fM [65]. Zhang et al. described the estimation of PSA in a sandwich-type electrochemical ELISA with fabrication of PtNP-ferrocenedicarboxylic acid based infinite coordination polymer (ICP) in combination with polyamidoamine dendrimers modified sensor electrode [66]. Xu et al. described the simultaneous detection of triple cancer biomarkers, namely PSA, CEA and AFP using newly developed carbon and gold (CGN) nanocomposite-based immunoprobes [67].

We mainly focus on the recent advances made by a various group in improvement strategies for electrochemical ELISA-based immunosensors for improving access to diagnostics, increased detection sensitivity and specificity, magnification of the signal, ease of handling, potential for automation and combination with miniaturized analytical systems, low cost and comparative simplicity for mass production. The development of generation and characterization of double nanobody-based sandwich ELISA for the detection of FGL1 in cancer patient serum [41]. San Martin et al., developed the "gold standard" ELISA-based electrochemical immunosensor for the

Biomarker	Clinical use	Cancer type	Specimen	Methodology
TP53	Diagnosis and monitoring of disease	Lymphocytic Leukemia/lymphoma	Serum	Immunoassay
RET	Diagnosis	NSCLC	Serum, plasma	Immunoassay
MS4A1(CD20 antigen)	Diagnosis and monitoring of disease	Non–Hodgkin's Lymphoma	Serum	Immunoassay
CCDC6-RET, KIF5B-RET, RET	Diagnosis	NSCLC	Serum, plasma	Immunoassay
PPP2R2A	Discriminating cancer from benign Disease	Prostate	Serum	Immunoassay
IGH	Diagnosis and monitoring of disease	Lymphocytic Leukemia/ lymphoma	Serum	Immunoassay
BRCA	Prediction of malignancy	Ovarian	Serum	Immunoassay
MS4A1 (CD20 antigen)	Prognosis, response to therapy	Lymphocytic Leukemia	Serum	Immunoassay
NPM1	Prognosis, response to therapy	Acute Myeloid Leukemia	Serum	Immunoassay
SSTR	Diagnosis and monitoring of disease	Gastroentero pancreatic neuroendocrine tumors	Serum, plasma	Immunoassay
HLA-A	Aid in differential diagnosis	Melanoma	Tissue	IHC
FIP1L1-PDGFRA	Diagnosis and monitoring of disease	Eosinophilic Leukemia	Serum	Immunoassay
MYD88	Prediction of malignancy	Macroglobulinemia	Serum	Immunoassay
FLT3	Diagnosis and monitoring of disease	Acute Myeloid Leukemia	Serum	Immunoassay
CD33	Diagnosis and monitoring of disease	Acute Myeloid Leukemia	Serum	Immunoassay
NECTIN4	Diagnosis and monitoring of disease	Metastatic Urothelial Cancer	Urine	Lateral flow immunoassay
IDH1, IDH2	Prognosis, response to therapy	Acute Myeloid Leukemia	Serum	Immunoassay
MYCN	Detection of tumors	Neuroblastoma	Serum	Immunoassay
IL2RA (CD25 antigen)	Prostate cancer diagnosis	T-cell lymphoma	Serum	Immunoassay

Biomarker	Clinical use	Cancer type	Specimen	Methodology
ROS1	Prognosis, response to therapy	NSCLC	Serum	Immunoassay
RAS	Prostate cancer diagnosis	Colorectal Cancer	Feces	Lateral flow immunoassay
MET	Prognosis, response to therapy	NSCLC	Serum	Immunoassay
TNFRSF8 (CD30)	progression of disease	T-cell lymphoma	Serum	Immunoassay
CD274 (PD-L1)	Aid in differential diagnosis	Merkel Cell Carcinoma NSCLC	Tissue	IHC
PDGFRA	Detection of tumors	Gastrointestinal Stromal Tumor	FFPE tissue	IHC
BRAF	Aid in differential diagnosis	Melanoma	Tissue	IHC
PML-RARA	Morphologic diagnosis of APL	Acute Promyelocytic leukemia	Serum	Immunoassay
PIK3CA	Prognosis, response to therapy	Breast	FFPE tissue	IHC
Anaplastic lymphoma kinase (ALK)	Prognosis, response to therapy for AL positive metastatic patients	NSCLC	Serum	Immunoassay
EGFR	Prognosis, response to therapy for EGFR mutation-positive patients	NSCLC	Serum	Immunoassay
Pro2PSA	Discriminating cancer from benign Disease	Prostate	Serum	Immunoassay
ROMA (HE4 + CA-125)	Prediction of malignancy	Ovarian	Serum	Immunoassay
OVA1 (multiple proteins)	Prediction of malignancy	Ovarian	Serum	Immunoassay
HE4	Monitoring recurrence or progression of disease	Ovarian	Serum	Immunoassay
Fibrin/fibrinogen degradation product (DR-70)	Monitoring progression of disease	Colorectal	Serum	Immunoassay
AFP-L3%	Risk assessment for development of disease	Hepatocellular	Serum	HPLC, microfluidic electrophoresis
Circulating Tumor Cells (EpCAM, CD45, cytokeratins 8, 18+, 19+)	Prediction of cancer progression and survival	Breast	Whole blood	Immunomagnetic fluorescence

Biomarker	Clinical use	Cancer type	Specimen	Methodology
p63 protein	Aid in differential diagnosis	Prostate	FFPE tissue	IHC
c-Kit	Detection of tumors, aid in selection of patients	Gastrointestinal stromal tumors	FFPE tissue	IHC
CA19–9	Monitoring disease status	Pancreatic	Serum, plasma	Immunoassay
Estrogen receptor (ER)	Prognosis, response to therapy	Breast	FFPE tissue	IHC
Progesterone receptor (PR)	Prognosis, response to therapy	Breast	FFPE tissue	IHC
HER-2/neu	Assessment for therapy	Breast	FFPE tissue	IHC
CA-125	Monitoring disease progression, response to therapy	Ovarian	Serum, plasma	Immunoassay
CA15–3	Monitoring disease response to therapy	Breast	Serum, plasma	Immunoassay
CA27.29	Monitoring disease response to therapy	Breast	Serum	Immunoassay
Free PSA	Discriminating cancer from benign Disease	Prostate	Serum	Immunoassay
Thyroglobulin	Aid in monitoring	Thyroid	Serum, plasma	Immunoassay
Nuclear Mitotic Apparatus protein	Diagnosis and monitoring of disease	Bladder	Urine	Lateral flow immunoassay
(NuMA, NMP22)	(professional and home use)			
Alpha-fetoprotein (AFP)[b]	Management of cancer	Testicular	Serum, plasma amniotic fluid[b]	Immunoassay
Total PSA	Prostate cancer diagnosis and Monitoring	Prostate	Serum	Immunoassay
Carcino-embryonic antigen	Aid in management and prognosis	Not specified	Serum, plasma	Immunoassay
Human hemoglobin (fecal occult blood)	Detection of fecal occult blood (home use)	Colorectal	Feces	Lateral flow immunoassay

Table 2.
List of FDA-approved protein tumor markers currently used in clinical practice.

single determination of both proteins PD-L1 and HIF-1α in terms of assay time, compatibility making their use suitable for untrained users at the point of attention [40]. The fabrication of sandwich ELISA type electrochemical aptasensor is developed for the instantaneous determination of two important biomarkers arcinoembryonic antigen (CEA) and cancer antigen 15–3 (CA 15–3) in breast cancer. CEA and CA 15–3 aptamers linked to gold nanoparticles/redox probe/graphene nanocomposite were used as biosensing probes for signal amplification and to enhance the sensitivity of the immunoassay [33]. Poly(ε-caprolactone) electrospun scaffolds (ePCL) are used to arrange for a microstructured substrate with a high surface-to-volume ratio, capable of binding E7 oncoproteins when used for enzyme-linked immunosorbent assay (ELISA) tests [31]. Interestingly, the ultrasensitive detection of cancer biomarker matrix metalloproteinase-1 in urine, saliva, bovine serum, and cell culture media of oral and brain cancers using label-free electrochemical immunosensor based on gold nanoparticle/ polyethyleneimine/reduced graphene oxide nanocomposites [29]. Li et al., 2021 constructed a simple label-free electrochemical immunosensor based on worm-like platinum with a sandwich-like structure [21]. The fabricated electrochemical immunosensor showed a wide linear range, enhanced detection limit, good selectivity and stability for the determination of alpha-fetoprotein.

Applications of Enzyme-linked Immunosorbent Assay in cancer research

- used to help diagnose certain diseases such as cancer
- For determination and quantification of tumor markers
- Detecting and measuring cell cycle checkpoint markers could be a promising pathway to better develop treatment strategies

2.2 Immunohistochemistry

Immunohistochemistry (IHC) is a fundamental method used for clinical decision making of diagnosis and prognosis of various cancers, such as breast [68], prostate [69], lung [70, 71]. It enables to find out the analysis of biomarker expression and tissue localization in cancer. Immunohistochemical techniques play critical roles a diagnosis and screening tools for distinguishing between malignant and benign with the help of biomarkers expression in lung cancer [72]. In recent years, the advancement of microfluidic-based immunohistochemistry represents clinically validated approaches to the standard chromogenic staining for rapid, accurate, and automated breast cancer diagnosis [73]. The automated chromogenic multiplexed immunohistochemistry assay approach provides an exclusive sample-sparing tool to characterize limited tissue samples in lung cancer and making it an emerging method in the clinical analysis for therapeutic decision making of advanced NSCLC, provided that validation in a larger population is performed. This implies limiting the number of tissue slides despite the existence of specific and sensitive biomarkers (ALK, ROS1, BRAF V600E, PD-L1) and the obligation to distinguish lung adenocarcinoma from squamous cell carcinoma [74].

Applications of Immunohistochemistry in cancer research

- To predict the prognosis of tumors by identification of enzymes, tumor-specific antigens, oncogenes, tumor suppressor genes, and tumor cell proliferation markers.

- To diagnosis of tumors of uncertain origin, primary as well as metastatic from unknown primary tumor.

- To predict therapeutic response in two important tumors, i.e. carcinoma of breast and prostate.

- Used to help tell the difference between different types of cancer

- To determine the function of specific gene products in fundamental biological processes such as development and apoptosis.

2.3 Flow cytometry

Flow cytometry is a versatile technique with applications in a variety of fields, including immunology, virology, molecular biology, cancer biology, and infectious disease surveillance. Flow cytometry is a technique for swiftly analysing single cells that are suspended in a buffered salt solution and flow through one or more lasers. Each cell is subjected to a visible light scattering analysis as well as one or more fluorescence parameters. Visible light scatter is assessed in two directions: forward (Forward Scatter or FSC), which shows the cell's relative size, and at 90° (Side Scatter or SSC), which reveals the cell's internal complexity or granularity. Fluorescence measurements are performed on samples that have been transfected and expressed with fluorescent proteins (for example, Green Fluorescent Protein, GFP), stained with fluorescent dyes (for example, Propidium Iodide, DNA), or stained with fluorescently attached antibodies (e.g., CD3 FITC). It enables simultaneous characterization of mixed populations of cells from blood and bone marrow as well as dissociable solid tissues such as lymph nodes, spleen, mucosal tissues, and solid malignancies.

The availability of new reagents has resulted in an explosion in the number of parameters utilised in flow cytometry investigations during the last several years. The number of fluorochromes used to conjugate monoclonal antibodies has increased dramatically, including tandem dyes and polymer dyes. Additionally, the number of fluorescent proteins accessible for transfection beyond GFP has increased, including mCherry, mBanana, mOrange, and mNeptune. These advancements in fluorochromes and technology have enabled tests with over 30 parameters to be performed. Data analysis is the final step of a flow cytometry experiment. The two-parameter histogram (dot plot) gating and analysis method is still widely utilised. However, as the number of factors and complexity of experiments expand, newer cluster data analysis techniques such as PCA, SPADE, and tSNE are being used. These enhanced data mining techniques enable the extraction of relevant information from the high-dimensional data generated by flow cytometry.

Since Mack Fulwyler initially invented the present kind of flow cytometers in 1965 [75], flow cytometry has now been used in quite a broad range of clinical areas for assessing protein expression in cancer cells [76, 77]. It is commonly used to diagnose of acute lymphoblastic leukemia [78]. Flow cytometry is a rapid and sensitive diagnostic method that makes it possible to characterize more satisfactorily the heterogeneous group of acute lymphoblastic leukemias. Flow cytometry has been historically used to detect the expression of CD56 in the diagnosis of chronic myelomonocytic leukemia (CMML). CD56 is a cell surface marker that presents the surface of monocytes [79]. Following over a decade of extensive research, high-throughput image-based flow cytometry is now an accepted and widely used tool in scientific research, particularly in the field of cancer biology. Many researchers have replaced microscopy-based clinical tools with image-based flow cytometry. Erber's team originally used image-based flow cytometry to identify the presence of promyelocytic leukemia (PML) bodies for the diagnosis and prognosis of acute myeloid leukemia (AML) [80].

Applications of Flow cytometry in cancer research

- Emerging as a tool for diagnosis of cancer (abnormal DNA content)
- Specific histopathological diagnosis (RNA for hematological cancers; surface markers for lymphoid and myeloid neoplasms)
- Prognosis (adverse impact of aneuploidy and high S percentage)
- Treatment (cytokinetically oriented, monoclonal antibodies, drug pharmacology)

2.4 Western blot

The Western blot (WB), also known as immunoblot, is an analytical and quantitative method for detecting particular proteins in various biological materials, including liquids and tissue/cellular homogenates [81]. Harry Towbin and colleagues developed the WB method in 1979. The WB approach provides clear and valuable information for assessing the phosphorylation state of a protein. We can evaluate the modified form of protein in the sample either qualitatively or quantitatively. Radenkovic et al., detected cyclin D1 expression in tumour and peritumoral tissue of breast cancer patients by Western blotting method to found that Cyclin D1 expression decreased significantly with each advanced clinical stage of disease and tumour size [25]. Kinase activity-tagged western blotting (KAT-WB) detected autophosphorylation of Tyr-kinase and site-specific phosphorylation by multiple kinases enables to interrogate multiple kinase signaling pathways without using radioactive substances [82]. Western blot analysis provides an opportunity to obtain more insight into cell cycle regulation factor in tumorigenesis, could spur the discovery of many more successful therapeutic targets [83].

Applications of Western blot in cancer research

- To detect the presence of cancer proteins biomarkers
- To determine the extent of post-translational modifications
- To verify protein expression in cloning applications
- To analyze protein and biomarker expression levels
- In antibody epitope mapping
- To test for markers of disease in clinical settings.

2.5 Mass spectrometry

The emphasis use of mass spectrometric analysis in clinical research has been on biomarker identification, which includes proteomics, lipidomics, and metabolomics [84, 85]. Metabolites, proteins and lipids have been shown to help distinguish between malignant and healthy tissue among the many compounds detectable with MS. To be identified as a qualifying biomarker, a molecule must be distinguishable from other molecules, ideally, the sample is simple, quick, and easy to collect, high sensitivity and specificity [86, 87], and is used to diagnose and prognosis for various cancer such as thyroid cancer [88], lung cancer [89, 90], bladder cancer [91], Pancreatic Cancer [92, 93], breast cancer [94], ovarian cancer [52], oral cancer [53], prostate cancer [95]. In general, various classes of molecules may function as biomarkers due to an imbalance of tumor-suppressing and promoting agents in cancer cells, regulating genetic changes, and changing the composition of lipids, metabolites, and proteins (**Figure 3**).

The MALDI-TOF MS combined with magnetic bead used for detecting serum protein biomarkers and establishment of boosting decision tree model for diagnosis of colorectal cancer patients [96], breast cancer [97]. These differentially regulated proteins were considered as potential biomarkers for the patients with CRC in the serum. The emerging mass spectrometry methods of nanoLC–MS/MS, targeted LC–MS/MS, and stable isotope-labeled multiple reactions monitoring (MRM) MS coupled to test

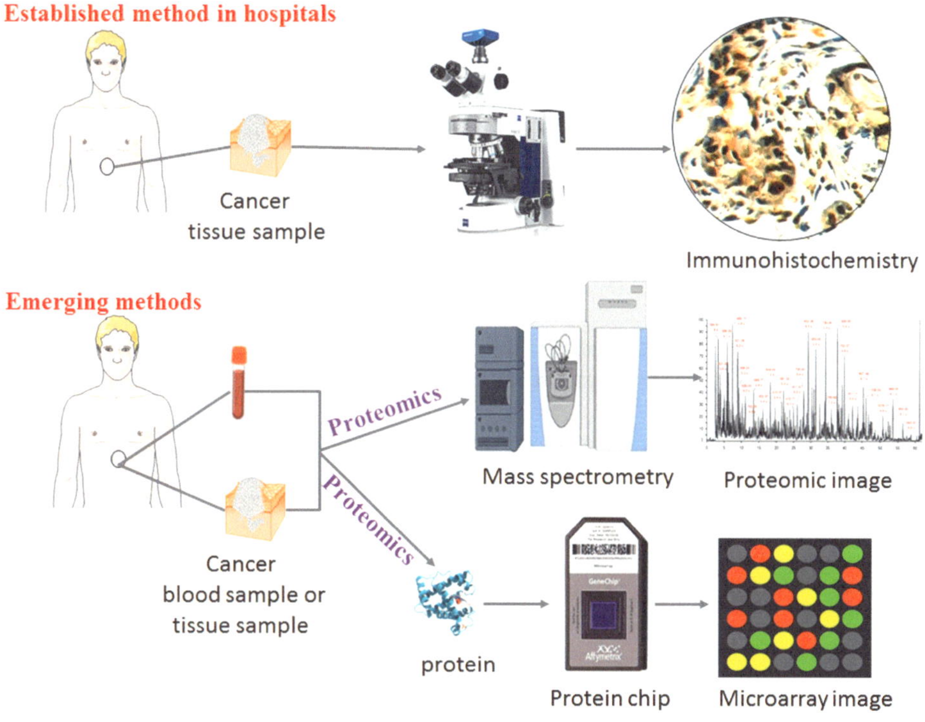

Figure 3.
Schematic representation showed that well established method immunohistochemistry was widely used in clinical diagnosis for cancer detection. In recent years, emerging proteomic technologies such as mass spectrometry and protein microarray has been developed for precise detection of cancer protein biomarkers.

machine learning algorithms and logistic regression used to analyze plasma samples from colorectal cancer patients. The novel peptide biomarkers combination of PF4, ITIH4, and APOE achieved sensitivity 84.5%, specificity 97.5% and an AUC of 0.96 in CRC diagnosis [98]. Moran et al., developed an intact protein assay to analyze PSA by capillary electrophoresis-electrospray ionization-mass spectrometry after affinity purification from prostate cancer patients' urine [99, 100].

The integrating mass spectrometry imaging and gold nanoparticle (AuNP)-based signal amplification was developed for quantitatively profiling protein biomarkers on the surface of exosomes in cancer diagnosis. Cancer protein biomarkers were modified with organic oligomers as mass tags and specific antibodies on AuNPs. Exosomes captured by the antibody-coated gold chip are recognized by the AuNPs probes, forming a sandwich immunoassay. Multiple protein biomarkers can be quantitatively detected from the exosomes with the mass tags by mass spectrometry imaging [101]. Park et al., 2019 developed a simple and robust cancer diagnostic method using matrix-assisted laser desorption/ionization time-of-flight mass spectrometry (MALDI-TOF MS)-based total serum protein fingerprinting to diagnose liver cancer [102]. The proteome profile of vimentin, tubulin beta 2C chain, tubulin alpha 1C chain, actin cytoplasmic 2, apolipoprotein A-I, and collagen alpha 2(VI) chain as a potential biomarker that exhibited differential expression in ovarian cancer using two-dimensional gel electrophoresis (2D-GE) and matrix-assisted laser desorption/ionization-time of flight mass spectrometric (MALDI-TOF MS) analysis [103]. The protein modifications in the N-glycosylation profile are usually associated with many cancers, like colorectal cancer. In turn, MALDI-TOF/MS and LC–MS methods are the most accurate technology in the quantification of N-glycans compositions in the serum and tissue of colorectal cancer patients [104]. The similar proteomic analysis was performed to identify potential lung cancer biomarkers such as CD5L, CLEC3B, ITIH4, SERFINF1, SAA4, SERFINC1, and C20ORF3 detected via a liquid biopsy-for the noninvasive diagnosis of lung cancer [105].

Applications of mass spectrometry in cancer research

- Discovery-based proteomic experiments with increasing cohort size is imperative for this technology to transfer to the clinic
- Application of tissue proteomics to cancer research is using it in a concerted effort to complement genomics.
- Proteomics biomarker research in human body fluids, urine and alternative liquid biopsies
- Mass spectrometry-based clinical proteomics in cancer biomarker research
- Prospective versus retrospective profiling of clinical specimens
- Human population heterogeneity & protein variability

2.6 Protein array

The analysis of protein biomarkers using these high throughput methods can provide robust and previously unachievable diagnostic and prognostic information for a variety of cancers [106, 107]. The analytical protein array is a useful developing technology that enables simultaneously analyzing >4000 protein samples in cell or tissues for biomarker discovery (Huang et al., 2017). The microarray is currently utilized to analyze biopsy samples in clinical research trials, essentially lead to the collection of information linked to posttranslational modifications of proteins reflecting the active status of signal pathways and networks [108]. This technique has the potential to enhance cancer detection, prognosis, and treatments. Protein microarray technology has been used effectively in fundamental and applied proteome research and affinity studies for protein identification, quantification, and functional analysis [109, 110]. A protein function array is made up of thousands of natural proteins that have been immobilised in a specific arrangement. When a functional protein array is used for serum protein profiling, autoantibodies are usually detected as biomarkers for diagnosis of cancer detection and for monitoring the cancer treatment due to their stability, specificity, and ease of detection, as compared with other serological components [111]. Protein microarrays have allowed researchers to examine functional protein dysregulation in various cancer namely colorectal cancers [112], pancreatic cancer [113]. Mirus et al., identified ERBB2, TNC and ESR1 in prediagnostic plasma from people that succumb to pancreatic ductal adenocarcinoma [114]. In addition to the understanding of the biological mechanisms, analytical protein arrays have also been applied to profile drug resistance [115].

In recent advancements in protein microarray is Reverse-phase protein array/microarray (RPPA/RPPM), which can precisely map functional proteomic profiling in individual cancer patient. The personalized therapy was prescribed by the identification of functional proteomics profilling. RPPA is an antibody-based highly quantitative proteomic technology, used for profiling the expression and modification of signaling proteins, mainly in low-abundance analytes cases. Clinical trials are using RPPA technology molecular-targeted therapeutics [116, 117]. Horton et al., 2021 found that the minimal effects on RPPA protein concentration distributions in peripheral blood and bone marrow, demonstrating that these preanalytical variables have been successfully managed in a multi-site clinical trial setting for leukemia [118]. A proteomic study was carried out for determining the levels of post-translational protein modifications and total protein expression in myeloproliferative neoplasms patients using RPPA [119]. These results highlight the robustness and the reproducibility of RPPA technology and its capacity to identify protein markers of cancer or response to therapy. Recent proteomics studies have focused on the expression of seven markers (CD5, CD10, BCL2, BCL6, MUM1, Ki-67, and C-MYC) is analyzed by RPPA using 37 diffuse large B-cell lymphomas (DLBCL) tissues [120]. These results suggest that RPPA could be applicable as a supportive tool for determining lymphoma prognosis. With all of these improvements, we believe that protein array technology will soon become a dominant tool for biomarker discovery in cancers.

Applications of protein array in cancer research

- Analytical and functional protein array for cancer biomarker discovery
- Personalized medicine in breast and ovarian cancers using protein microarray
- Protein profiling of cancer cells using protein microarray
- Drug discovery for target identification and validation

2.7 Microfluidics

Microfluidic technology, as new creativity has a great impact on automation and miniaturization via handling a small volume of materials and samples for cancer diagnosis [121]. This method has be considered as a primary screening tool for diagnosing breast cancer based on its robustness, high throughput, low energy requirements, excellent accuracy and accessibility to the general public [122]. A miniaturized instrument was developed for chemiluminescence detection and signal analysis with the advances in microfluidic technology. The system was validated by testing four biomarkers of colorectal cancer using plasma samples from patients [123]. Another design of the microfluidic device, magnetic nanoparticles (Fe_3O_4NPs) was successfully functionalized with an exosome-binding antibody (anti-CD9) to mediate the magnetic capture in a microdevice. The captured exosomes were then subjected to analysis of CA19–9, a protein often used to monitor pancreatic cancer patients [124]. In the line of discovery, the nuclear matrix protein 22 (NMP22) and bladder cancer antigen (BTA) from the urine samples was detected using the micro-fluidic paper-based analytical device (mu PAD) was developed by Jiang [125]. This method is feasible for home-based self-detection from urine samples within 10 min for the total process, which provides a new way for quick, economical, and convenient tumor diagnosis, prognosis evaluation, and drug response (**Figure 4**). The functions and recent development of microfluidic chip to provide great potential for advancing noninvasive cancer diagnosis [126, 127].

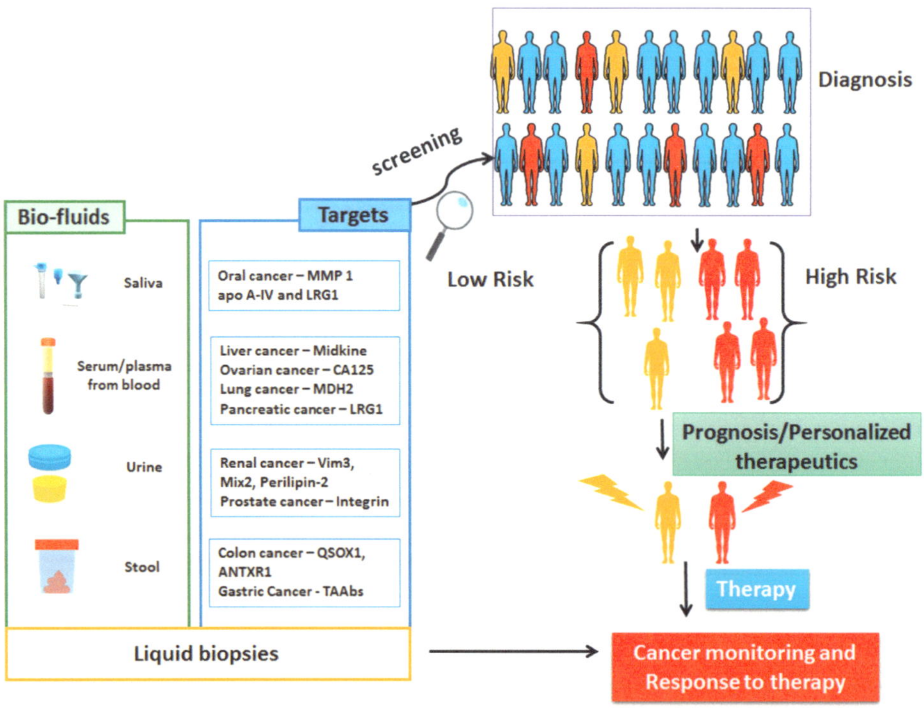

Figure 4.
Microfluidic technology, as new creativity has a great impact on automation and miniaturization via handling a small volume of materials and samples for cancer diagnosis. An effective management of cancer diagnosis screening by using body fluids and cancer protein biomarkers for diagnosis, prognosis, therapy and monitoring to treatment.

Microfluidic devices which used to mimic cancer metastasis process are usually applied to several cell types in order to culture two or more organoids. Different organoids are separated by some specific biomaterials, such as polydimethylsiloxane (PDMS), and connected with each other by channels and controllable fluids. Xu et al. designed and constructed a multi-organ microfluidic chip to mimic lung cancer metastasis to the brain, bone and liver. In this platform, organoids were divided into different chambers, including upstream lung organoid and three downstream organoids. Different types of cells were seeded in each chamber to culture different organoids and each organoid were linked by side channels. The culture medium flowed through microvascular channels to simulate blood circulation. At the same time, a circulating vacuum was applied to mimic the physiological breathing [128]. This system provided a physiologically relevant context to recapitulate the complex process of lung cancer metastasis and help us to effectively explore the underlying mechanism of lung cancer metastasis.

The integration between 3D bioprinting and microfluidic chip has given microfluidic chip greater potential to model cancers. Traditionally, in cancer modeling on chip, microfabrication such as micromachining, photolithography and injection molding, are used in the fabrication of microfluidic chips [129]. These methods have high resolution and accuracy, but their high cost, complex process and difficult reproducibility greatly limited the development of microfluidic chip [130]. The emerging of 3D printing technology greatly simplifies the fabrication process of microfluidic chips. The biggest characteristic of microfluidic chip is the customizability, which means microfluidic chip is a very flexible scientific tool that can accommodate with advanced technologies. To date, microfluidic chip shows tremendous promise in cancer diagnosis and treatment. Microfluidic chip can be applied in everything from anticancer drug development and screening to cancer modeling and diagnosis.

Applications of Microfluidics in cancer research

- The development of cancer preclinical model; Animal models, 2D culture, 3D culture, as well as tumor organoid.
- Detection of cancer biomarkers
- Anti-cancer drug screening and nano-drug preparation
- Exploring tumor heterogeneity on microfluidic chip

3. Conclusions

Over the past decade, protein profiling has emerged as a dynamic discipline, capable of generating a comprehensive perspective of protein patterns, modification of protein in various tissue-specific cancer types and mechanisms of cancer progression. Proteomics studies and analytical techniques are developed with modern materials for precise detection of tumor-specific alteration in proteins. Recent times, the proteomics technologies have fabricated with modern nanocomposites provide a new and more efficient methods for protein detection and identifying biomarkers for the early detection of cancer. The upgraded proteomics technology will modify the current pathological classification and grading methods of cancer during the next decade. Proteomic technologies will have an impact on the diagnosis and management of cancer.

Author details

Shenbagamoorthy Sundarraj[1*], Gopalan Rajagopal[1], Balaji Sundaramahalingam[1], Madasamy Sundar[2] and Ramar Thangam[3]

1 Postgraduate and Research Department of Zoology, Ayya Nadar Janaki Ammal College, Sivakasi, Tamil Nadu, India

2 Centre for Research and Postgraduate Studies in Botany, Ayya Nadar Janaki Ammal College, Sivakasi, Tamil Nadu, India

3 Department of Materials Science and Engineering, Korea University, Seoul, Republic of Korea

*Address all correspondence to: sundarrajbu09@gmail.com

References

[1] Henley SJ, Ward EM, Scott S, Ma J, Anderson RN, Firth AU, et al. Annual report to the nation on the status of cancer, part I: National cancer statistics. Cancer. 2020;126(10):2225-2249.

[2] de Martel C, Georges D, Bray F, Ferlay J, Clifford GM. Global burden of cancer attributable to infections in 2018: a worldwide incidence analysis. The Lancet Global Health. 2020;8(2): e180–e190.

[3] Guo R, Pan C, Shen J, Liu C. New serum biomarkers for detection of esophageal carcinoma using Matrix-assisted laser desorption/ionization time-of-flight mass spectrometry. Journal of cancer research and clinical oncology. 2011;137(3):513-519.

[4] K Nowicki T, Markiet K, Szurowska E. Diagnostic imaging of hepatocellular carcinoma-A pictorial essay. Current Medical Imaging. 2017;13(2):140-153.

[5] Løberg M, Lousdal ML, Bretthauer M, Kalager M. Benefits and harms of mammography screening. Breast Cancer Research. 2015;17(1):1-12.

[6] Meric-Bernstam F, Mills GB. Overcoming implementation challenges of personalized cancer therapy. Nature reviews Clinical oncology. 2012;9(9): 542-548.

[7] Hing JX, Mok CW, Tan PT, Sudhakar SS, Seah CM, Lee WP, et al. Clinical utility of tumour marker velocity of cancer antigen 15-3 (CA 15-3) and carcinoembryonic antigen (CEA) in breast cancer surveillance. The Breast. 2020;52:95-101.

[8] Kim DW, Chen M, Wu J, Huland H, Graefen M, Tilki D, et al. Prostate-specific antigen levels of ≤ 4 and > 4 ng/mL and risk of prostate cancer–specific mortality in men with biopsy Gleason score 9 to 10 prostate cancer. Cancer. 2021;

[9] Song B, Wang L, Zhang Y, Li N, Dai H, Xu H, et al. Combined detection of HER2, Ki67, and GSTP1 genes on the diagnosis and prognosis of breast cancer. Cancer biotherapy & radiopharmaceuticals. 2019;34(2):85-90.

[10] Jayasekara H, MacInnis RJ, Chamberlain JA, Dite GS, Leoce NM, Dowty JG, et al. Mortality after breast cancer as a function of time since diagnosis by estrogen receptor status and age at diagnosis. International journal of cancer. 2019;145(12):3207-3217.

[11] Tainsky MA. Genomic and proteomic biomarkers for cancer: a multitude of opportunities. Biochimica et Biophysica Acta (BBA)-Reviews on Cancer. 2009;1796(2):176-193.

[12] Sokolenko AP, Imyanitov EN. Molecular diagnostics in clinical oncology. Frontiers in molecular biosciences. 2018;5:76.

[13] Zhang X, Sun X-F, Shen B, Zhang H. Potential applications of DNA, RNA and protein biomarkers in diagnosis, therapy and prognosis for colorectal cancer: a study from databases to AI-assisted verification. Cancers. 2019;11(2):172.

[14] Duffy MJ. Tumor markers in clinical practice: a review focusing on common solid cancers. Medical Principles and Practice. 2013;22(1):4-11.

[15] Enroth S, Berggrund M, Lycke M, Broberg J, Lundberg M, Assarsson E, et al. High throughput proteomics

identifies a high-accuracy 11 plasma protein biomarker signature for ovarian cancer. Communications biology. 2019;2(1):1-12.

[16] Henderson MC, Hollingsworth AB, Gordon K, Silver M, Mulpuri R, Letsios E, et al. Integration of serum protein biomarker and tumor associated autoantibody expression data increases the ability of a blood-based proteomic assay to identify breast cancer. PLoS One. 2016;11(8):e0157692.

[17] Lin D, Feng S, Pan J, Chen Y, Lin J, Chen G, et al. Colorectal cancer detection by gold nanoparticle based surface-enhanced Raman spectroscopy of blood serum and statistical analysis. Optics express. 2011;19(14):13565-13577.

[18] Li W, Moore MJ, Vasilieva N, Sui J, Wong SK, Berne MA, et al. Angiotensin-converting enzyme 2 is a functional receptor for the SARS coronavirus. Nature. 2003;426(6965):450-454.

[19] Wong Y-L, Ramanathan A, Yuen KM, Mustafa WMW, Abraham MT, Tay KK, et al. Comparative sera proteomics analysis of differentially expressed proteins in oral squamous cell carcinoma. PeerJ. 2021;9:e11548.

[20] El-Shayeb AF, El-Habachi NM, Mansour AR, Zaghloul MS. Serum midkine is a more sensitive predictor for hepatocellular carcinoma than Dickkopf-1 and alpha-L-fucosidase in cirrhotic HCV patients. Medicine. 2021;100(17).

[21] Li S-S, Tan Y-Y, Zhang Y, Liu M, Liu A. A simple electrochemical immunosensor based on worm-like platinum for highly sensitive determination of alpha-fetoprotein. Bioelectrochemistry. 2021;140:107804.

[22] Guo X, Zhang Z, Yang M, Bing P, Yan X, Yang Q, et al. Time-Frequency Double Domain Resolving by Electromagnetically Induced Transparency Metasensors for Rapid and Label-Free Detection of Cancer Biomarker Midkine. Optics and Lasers in Engineering. 2021;142:106566.

[23] Chaudhary MK, Jaiswal R, Singh N, Pandey A, Ali W. Evaluation of Serum s EGFR in Ovarian Tumors, its Comparison with Serum CA125, HE4 and Correlation with Histopathological Types. Indian Journal of Gynecologic Oncology. 2021;19(2):1-6.

[24] von Brandenstein M, Herden J, Köditz B, Huerta M, Nestler T, Heidenreich A, et al. Non-invasive urine markers for the differentiation between RCCs and oncocytoma. Journal of clinical laboratory analysis. 2021;35(5): e23762.

[25] Radenkovic S, Konjevic G, Nikitovic M, Stojanovic-Rundic S, Plesinac-Karapandzic V, Milovic-Kovacevic M, et al. Evaluation of Cyclin D1 expression by western blotting methods and immunohistochemistry in breast cancer patients. Journal of BU ON: Official Journal of the Balkan Union of Oncology. 2021;26(2):475-482.

[26] Liang H, Dong J, Cheng Z, Li Q, Feng D, Ling B. B-cell receptor-associated protein 31 promotes migration and invasion in ovarian cancer cells. Experimental and Therapeutic Medicine. 2021;22(2):1-12.

[27] Ma Y, Tian P, Chen Z, Yue D, Liu C, Li C, et al. Urinary malate dehydrogenase 2 is a new biomarker for early detection of non-small-cell lung cancer. Cancer Science. 2021;112(6):2349.

[28] Xu Z, Hu Y, Yu Z. Effect of the ACY-1 gene on HER2 and TRAIL expression in rectal carcinoma. Experimental and Therapeutic Medicine. 2021;22(2):1-7.

[29] Liu X, Lin L-Y, Tseng F-Y, Tan Y-C, Li J, Feng L, et al. Label-free electrochemical immunosensor based on gold nanoparticle/polyethyleneimine/reduced graphene oxide nanocomposites for the ultrasensitive detection of cancer biomarker matrix metalloproteinase-1. Analyst. 2021;

[30] Jiang D, Zhang X, Liu M, Wang Y, Wang T, Pei L, et al. Discovering panel of autoantibodies for early detection of lung cancer based on focused protein array. Frontiers in immunology. 2021;12:710.

[31] Falcucci S, Paolini F, Mileo AM, Franconi R, Massa S, Rinaldi A, et al. ePCL Electrospun Microfibrous Layers for Immune Assays: Sensitive ELISA for the Detection of Serum Antibodies Against HPV16 E7 Oncoprotein. ACS omega. 2021;6(13):8778-8783.

[32] Nakaguro M, Tanigawa M, Hirai H, Yamamoto Y, Urano M, Takahashi RH, et al. The Diagnostic Utility of RAS Q61R Mutation-specific Immunohistochemistry in Epithelial-Myoepithelial Carcinoma. The American journal of surgical pathology. 2021;45(7): 885-894.

[33] Shekari Z, Zare HR, Falahati A. Dual assaying of breast cancer biomarkers by using a sandwich–type electrochemical aptasensor based on a gold nanoparticles–3D graphene hydrogel nanocomposite and redox probes labeled aptamers. Sensors and Actuators B: Chemical. 2021;332:129515.

[34] Zemskova MY, Marinets M V, Sivkov A V, Pavlova J V, Shibaev AN, Sorokin KS. Integrin Alpha V in Urine: A Novel Noninvasive Marker for Prostate Cancer Detection. Frontiers in oncology. 2020;10.

[35] Lin Z, Süsskind D. Exploring the role of BAFF as biomarker in the detection of uveal melanoma metastases. Journal of Cancer Research and Clinical Oncology. 2021;147(5):1389-1405.

[36] Ganig N, Baenke F, Thepkaysone M-L, Lin K, Rao VS, Wong FC, et al. Proteomic Analyses of Fibroblast-and Serum-Derived Exosomes Identify QSOX1 as a Marker for Non-invasive Detection of Colorectal Cancer. Cancers. 2021;13(6):1351.

[37] Choi YJ, Yoon W, Lee A, Han Y, Byun Y, Kang JS, et al. Diagnostic model for pancreatic cancer using a multi-biomarker panel. Annals of surgical treatment and research. 2021;100(3):144.

[38] Zhu Y-L, Lian Y-M, Wang J-K, Chen Z-P, Yu R-Q. Ultrasensitive detection of protein biomarkers by MALDI-TOF mass spectrometry based on $ZnFe2O4$ nanoparticles and mass tagging signal amplification. Talanta. 2021;224:121848.

[39] Cui C, Duan Y, Qiu C, Wang P, Sun G, Ye H, et al. Identification of Novel Autoantibodies Based on the Human Proteomic Chips and Evaluation of Their Performance in the Detection of Gastric Cancer. Frontiers in oncology. 2021;11:342.

[40] Muñoz-San Martín C, Gamella M, Pedrero M, Montero-Calle A, Pérez-Ginés V, Camps J, et al. Anticipating metastasis through electrochemical immunosensing of tumor hypoxia biomarkers. Analytical and bioanalytical chemistry. 2021;1-14.

[41] Zhang W, Liu T, Wu M, Chen X, Han L, Shi Z, et al. Development of a nanobody-based immunoassay for the sensitive detection of fibrinogen-like protein 1. Acta Pharmacologica Sinica. 2021;1-9.

[42] Leung SOA, Feldman S, Kalyanaraman R, Shanmugam V,

Worley MJ, Berkowitz RS, et al. Triaging abnormal cervical cancer screening tests using p16INK4a detection by ELISA on fresh cervical samples. American Journal of Reproductive Immunology. 2021;e13394.

[43] Hao Y, Ye M, Chen X, Zhao H, Hasim A, Guo X. Discovery and validation of FBLN1 and ANT3 as potential biomarkers for early detection of cervical cancer. Cancer cell international. 2021;21(1):1-13.

[44] Chen J, Li H, Xu W, Guo X. Evaluation of serum ATX and LPA as potential diagnostic biomarkers in patients with pancreatic cancer. BMC gastroenterology. 2021;21(1):1-10.

[45] Pietrzyk Ł, Korolczuk A, Matysek M, Arciszewski MB, Torres K. Clinical Value of Detecting Tumor Endothelial Marker 8 (ANTXR1) as a Biomarker in the Diagnosis and Prognosis of Colorectal Cancer. Cancer Management and Research. 2021;13:3113.

[46] Yang HX, Yang Y, Li XD, Miao XM, Yang C, Zhi DF, et al. The detection of SAS1B in serum provides clues for early diagnosis of thyroid cancer. Eur Rev Med Pharmacol Sci. 2021;25(3):1410-1424.

[47] Lokman NA, Ricciardelli C, Stephens AN, Jobling TW, Hoffmann P, Oehler MK. Diagnostic Value of Plasma Annexin A2 in Early-Stage High-Grade Serous Ovarian Cancer. Diagnostics 2021, 11, 69. s Note: MDPI stays neutral with regard to jurisdictional claims in ...; 2021.

[48] Nie X-C, He F, Lan C, Niu J-M, Xia P. Combined Serum DKK3 and Circulating CD133 Cells as Prognostic Biomarkers for Ovarian Cancer Patients. OncoTargets and therapy. 2021;14:427.

[49] Sharafeldin M, Chen T, Ozkaya GU, Choudhary D, Molinolo AA, Gutkind JS, et al. Detecting cancer metastasis and accompanying protein biomarkers at single cell levels using a 3D-printed microfluidic immunoarray. Biosensors and Bioelectronics. 2021;171:112681.

[50] Thangsunan P, Lal N, Tiede C, Moul S, Robinson JI, Knowles MA, et al. Affimer-based impedimetric biosensors for fibroblast growth factor receptor 3 (FGFR3): a novel tool for detection and surveillance of recurrent bladder cancer. Sensors and Actuators B: Chemical. 2021;326:128829.

[51] Mawaribuchi S, Haramoto Y, Tateno H, Onuma Y, Aiki Y, Ito Y. rBC2LCN lectin as a potential probe of early-stage HER2-positive breast carcinoma. FEBS open bio. 2020;10(6): 1056-1064.

[52] Schuster-Little N, Madera S, Whelan R. Developing a mass spectrometry–based assay for the ovarian cancer biomarker CA125 (MUC16) using suspension trapping (STrap). Analytical and bioanalytical chemistry. 2020;412(24):6361-6370.

[53] Hsiao Y-C, Lin S-Y, Chien K-Y, Chen S-F, Wu C-C, Chang Y-T, et al. An immuno-MALDI mass spectrometry assay for the oral cancer biomarker, matrix metalloproteinase-1, in dried saliva spot samples. Analytica chimica acta. 2020;1100:118-130.

[54] Hu R, Gupta R, Wang Z, Wang C, Sun H, Singamaneni S, et al. Bioplasmonic paper–based assay for perilipin-2 non-invasively detects renal cancer. Kidney international. 2019;96(6):1417-1421.

[55] Chang S-C, Lin W-L, Chang Y-F, Lee C-T, Wu J-S, Hsu P-H, et al. Glycoproteomic identification of novel plasma biomarkers for oral cancer. Journal of food and drug analysis. 2019;27(2):483-493.

[56] Šuster NK, Meznaric M, Škorja N, Virant-Klun I, Verdenik I, Smrkolj Š. Cancer stem cell-related marker NANOG expression in ovarian serous tumors using Western blotting and immunohistochemistry: comparison of two techniques. European Journal of Gynaecological Oncology. 2019;40(6): 948-952.

[57] Kotzsch M, Kirchner T, Soelch S, Schäfer S, Friedrich K, Baretton G, et al. Inverse association of rab31 and mucin-1 (CA15-3) antigen levels in estrogen receptor-positive (ER+) breast cancer tissues with clinicopathological parameters and patients' prognosis. American journal of cancer research. 2017;7(9):1959.

[58] Miyashita M, Gonda K, Tada H, Watanabe M, Kitamura N, Kamei T, et al. Quantitative diagnosis of HER 2 protein expressing breast cancer by single-particle quantum dot imaging. Cancer medicine. 2016;5(10):2813-2824.

[59] Chikkaveeraiah B V, Mani V, Patel V, Gutkind JS, Rusling JF. Microfluidic electrochemical immunoarray for ultrasensitive detection of two cancer biomarker proteins in serum. Biosensors and Bioelectronics. 2011;26(11): 4477-4483.

[60] Clark MF, Adams AN. Characteristics of the microplate method of enzyme-linked immunosorbent assay for the detection of plant viruses. Journal of general virology. 1977;34(3):475-483.

[61] Engvall E, Perlmann P. Enzyme-linked immunosorbent assay (ELISA) quantitative assay of immunoglobulin G. Immunochemistry. 1971;8(9):871-874.

[62] Reen DJ. Enzyme-linked immunosorbent assay (ELISA). Basic Protein and Peptide Protocols. 1994; 461-466.

[63] Oesterling JE. Prostate specific antigen: a critical assessment of the most useful tumor marker for adenocarcinoma of the prostate. The Journal of urology. 1991;145(5):907-923.

[64] Freedland SJ, Humphreys EB, Mangold LA, Eisenberger M, Dorey FJ, Walsh PC, et al. Risk of prostate cancer–specific mortality following biochemical recurrence after radical prostatectomy. Jama. 2005;294(4):433-439.

[65] Shim JU, Ranasinghe RT, Smith CA, Ibrahim SM, Hollfelder F, Huck WTS, et al. Ultrarapid generation of femtoliter microfluidic droplets for single-molecule-counting immunoassays. ACS Nano 7, 5955-5964. 2013.

[66] Zhang B, Liu B, Chen G, Tang D. Redox and catalysis 'all-in-one'infinite coordination polymer for electrochemical immunosensor of tumor markers. Biosensors and Bioelectronics. 2015;64:6-12.

[67] Xu T, Liu N, Yuan J, Ma Z. Triple tumor markers assay based on carbon–gold nanocomposite. Biosensors and Bioelectronics. 2015;70:161-166.

[68] Bonacho T, Rodrigues F, Liberal J. Immunohistochemistry for diagnosis and prognosis of breast cancer: a review. Biotechnic & Histochemistry. 2020; 95(2): 71-91.

[69] Mandel P, Wenzel M, Hoeh B, Welte MN, Preisser F, Inam T, et al. Immunohistochemistry for Prostate Biopsy—Impact on Histological Prostate Cancer Diagnoses and Clinical Decision Making. Current Oncology. 2021;28(3): 2123-2133.

[70] Albero-González R, Munné-Collado J, Pijuan L, Simón M, Gimeno-Beltrán J, Mojal S, et al. Complementary value of electron microscopy and immunohisto-chemistry in the diagnosis of non-small

cell lung cancer: A potential role for electron microscopy in the era of targeted therapy. Ultrastructural pathology. 2019;43(6):237-247.

[71] Imai H, Wasamoto S, Yamaguchi O, Suzuki K, Sugiyama T, Uchino J, et al. Efficacy and safety of first-line pembrolizumab monotherapy in elderly patients (aged≥ 75 years) with non-small cell lung cancer. Journal of cancer research and clinical oncology. 2020; 146(2):457-466.

[72] Inamura K. Update on immunohistochemistry for the diagnosis of lung cancer. Cancers. 2018;10(3):72.

[73] Aimi F, Procopio M-G, Flores MTA, Brouland J-P, Piazzon N, Brajkovic S, et al. Microfluidic-based immunohisto-chemistry for breast cancer diagnosis: a comparative clinical study. Virchows Archiv. 2019;475(3):313-323.

[74] Ilie M, Hofman V, Ortholan C, Bonnetaud C, Coëlle C, Mouroux J, et al. Predictive clinical outcome of the intratumoral CD66b-positive neutrophil-to-CD8-positive T-cell ratio in patients with resectable nonsmall cell lung cancer. Cancer. 2012;118(6):1726-1737.

[75] Fulwyler MJ. Electronic separation of biological cells by volume. Science. 1965;150(3698):910-911.

[76] Orfao A, Ciudad J, Gonzalez M, Lopez A, Del Mar Abad M, Paz Bouza JI, et al. Flow cytometry in the diagnosis of cancer. Scandinavian Journal of Clinical and Laboratory Investigation. 1995;55 (sup221):145-52.

[77] Danova M, Torchio M, Comolli G, Sbrana A, Antonuzzo A, Mazzini G. The role of automated cytometry in the new era of cancer immunotherapy. Molecular and clinical oncology. 2018;9(4):355-361.

[78] Starý J, Sokol L, Hausner P, Hrodek O, Sedmihorská J, Volejníková J, et al. Use of flow cytometry in the diagnosis of acute leukemias in childhood. Casopis lekaru ceskych. 1992;131(12):364-367.

[79] Hudson CA, Burack WR, Bennett JM. Emerging utility of flow cytometry in the diagnosis of chronic myelomonocytic leukemia. Leukemia research. 2018;73:12-15.

[80] Grimwade L, Gudgin E, Bloxham D, Scott MA, Erber WN. PML protein analysis using imaging flow cytometry. Journal of clinical pathology. 2011;64(5):447-450.

[81] Ghosh R, Gilda JE, Gomes A V. The necessity of and strategies for improving confidence in the accuracy of western blots. Expert review of proteomics. 2014;11(5):549-560.

[82] Eto M, Katsuki S, Tanaka Y, Takeya K. Kinase activity-tagged western blotting assay. BioTechniques. 2019;68(4):211-213.

[83] Li L-H, Huang Q-M, Barbero M, Liu L, Nguyen T-T, Beretta-Piccoli M, et al. Quantitative proteomics analysis to identify biomarkers of chronic myofascial pain and therapeutic targets of dry needling in a rat model of myofascial trigger points. Journal of pain research. 2019;12:283.

[84] Zhou W, Petricoin EF, Longo C. Mass spectrometry-based biomarker discovery. In: Molecular Profiling. Springer; 2017. p. 297-311.

[85] Steffen P, Kwiatkowski M, Robertson WD, Zarrine-Afsar A, Deterra D, Richter V, et al. Protein species as diagnostic markers. Journal of proteomics. 2016;134:5-18.

[86] Brower V. Biomarkers: Portents of malignancy. Nature. 2011;471(7339): S19–S20.

[87] Rosenling T, Stoop MP, Smolinska A, Muilwijk B, Coulier L, Shi S, et al. The impact of delayed storage on the measured proteome and metabolome of human cerebrospinal fluid. Clinical chemistry. 2011;57(12):1703-1711.

[88] Li F, Pan D, He Y, Wu Y, Peng J, Li J, et al. Using ultrasound features and radiomics analysis to predict lymph node metastasis in patients with thyroid cancer. BMC surgery. 2020;20(1):1-6.

[89] Wu H, Pan Y, Kopylov AT, Zgoda V, Ma M, Wang C, et al. Assessment of Serological Early Biomarker Candidates for Lung Adenocarcinoma by using Multiple Reaction Monitoring-Mass Spectrometry. PROTEOMICS–Clinical Applications. 2020;14(4):1900095.

[90] Zhou Y, Wu C, Lu G, Hu Z, Chen Q, Du X. FGF/FGFR signaling pathway involved resistance in various cancer types. Journal of Cancer. 2020;11(8): 2000.

[91] Nedjadi T, Benabdelkamal H, Albarakati N, Masood A, Al-Sayyad A, Alfadda AA, et al. Circulating proteomic signature for detection of biomarkers in bladder cancer patients. Scientific reports. 2020;10(1):1-10.

[92] Jia K, Zhao X, Dang X. Mass spectrometry-based iTRAQ analysis of serum markers in patients with pancreatic cancer. Oncology letters. 2020;19(6):4106-4114.

[93] Kim Y, Yeo I, Huh I, Kim J, Han D, Jang J-Y, et al. Development and Multiple Validation of the Protein Multi-marker Panel for Diagnosis of Pancreatic Cancer. Clinical Cancer Research. 2021;27(8):2236-2245.

[94] Lee SB, Bose S, Ahn SH, Son BH, Ko BS, Kim HJ, et al. Breast cancer diagnosis by analysis of serum N-glycans using MALDI-TOF mass spectroscopy. Plos one. 2020;15(4):e0231004.

[95] Geisler C, Gaisa NT, Pfister D, Fuessel S, Kristiansen G, Braunschweig T, et al. Identification and validation of potential new biomarkers for prostate cancer diagnosis and prognosis using 2D-DIGE and MS. BioMed research international. 2015;2015.

[96] Liu C, Pan C, Shen J, Wang H, Yong L. MALDI-TOF MS combined with magnetic beads for detecting serum protein biomarkers and establishment of boosting decision tree model for diagnosis of colorectal cancer. International journal of medical sciences. 2011;8(1):39.

[97] de Noo ME, Deelder A, van der Werff M, Özalp A, Mertens B, Tollenaar R. MALDI-TOF serum protein profiling for the detection of breast cancer. Oncology Research and Treatment. 2006;29(11):501-506.

[98] Tsai I, Su EC-Y, Tsai I-L, Lin C-Y. Clinical assay for the early detection of colorectal cancer using mass spectrometric wheat germ agglutinin multiple reaction monitoring. Cancers. 2021;13(9):2190.

[99] Otto JJ, Correll VL, Engstroem HA, Hitefield NL, Main BP, Albracht B, et al. Targeted Mass Spectrometry of a Clinically Relevant PSA Variant from Post-DRE Urines for Quantitation and Genotype Determination. PROTEOMICS–Clinical Applications. 2020;14(6):2000012.

[100] Moran AB, Domínguez-Vega E, Nouta J, Pongracz T, de Reijke TM, Wuhrer M, et al. Profiling the proteoforms of urinary prostate-specific

antigen by capillary electrophoresis–mass spectrometry. Journal of Proteomics. 2021;238:104148.

[101] Wang Y, Zhang K, Huang X, Qiao L, Liu B. Mass spectrometry imaging of mass tag immunoassay enables the quantitative profiling of biomarkers from dozens of exosomes. Analytical Chemistry. 2020;93(2):709-714.

[102] Park H-G, Jang K-S, Park H-M, Song W-S, Jeong Y-Y, Ahn D-H, et al. MALDI-TOF MS-based total serum protein fingerprinting for liver cancer diagnosis. Analyst. 2019;144(7):2231-2238.

[103] Noreen S, Akhtar S, Batool T, Gardner QA, Akhtar MW. Tubulin Beta 2C Chain (TBB2C), a Potential Marker of Ovarian Cancer, an Insight from Ovarian Cancer Proteome Profile. ACS omega. 2021;6(16):10506-10514.

[104] Coura M de, Barbosa EA, Brand GD, Bloch C, de Sousa JB. Identification of Differential N-Glycan Compositions in the Serum and Tissue of Colon Cancer Patients by Mass Spectrometry. Biology. 2021;10(4):343.

[105] Choi E-S, Faruque H Al, Kim J-H, Kim KJ, Choi JE, Kim BA, et al. CD5L as an Extracellular Vesicle-Derived Biomarker for Liquid Biopsy of Lung Cancer. Diagnostics. 2021;11(4):620.

[106] Grant GM, Fortney A, Gorreta F, Estep M, Del Giacco L, Van Meter A, et al. Microarrays in cancer research. Anticancer research. 2004;24(2A): 441-448.

[107] Russo G, Zegar C, Giordano A. Advantages and limitations of microarray technology in human cancer. Oncogene. 2003;22(42):6497-6507.

[108] Liotta LA, Espina V, Mehta AI, Calvert V, Rosenblatt K, Geho D, et al. Protein microarrays: meeting analytical challenges for clinical applications. Cancer cell. 2003;3(4):317-325.

[109] Poetz O, Schwenk JM, Kramer S, Stoll D, Templin MF, Joos TO. Protein microarrays: catching the proteome. Mechanisms of ageing and development. 2005;126(1):161-170.

[110] Talapatra A, Rouse R, Hardiman G. Protein microarrays: challenges and promises. Pharmacogenomics. 2002;3(4):527-536.

[111] Heo C-K, Bahk YY, Cho E-W. Tumor-associated autoantibodies as diagnostic and prognostic biomarkers. BMB reports. 2012;45(12):677.

[112] Barderas R, Babel I, Diaz-Uriarte R, Moreno V, Suarez A, Bonilla F, et al. An optimized predictor panel for colorectal cancer diagnosis based on the combination of tumor-associated antigens obtained from protein and phage microarrays. Journal of proteomics. 2012;75(15):4647-4655.

[113] Gerdtsson AS, Malats N, Säll A, Real FX, Porta M, Skoog P, et al. A multicenter trial defining a serum protein signature associated with pancreatic ductal adenocarcinoma. International journal of proteomics. 2015;2015.

[114] Mirus JE, Zhang Y, Li CI, Lokshin AE, Prentice RL, Hingorani SR, et al. Cross-species antibody microarray interrogation identifies a 3-protein panel of plasma biomarkers for early diagnosis of pancreas cancer. Clinical Cancer Research. 2015;21(7):1764-1771.

[115] Smith L, Watson MB, O'Kane SL, Drew PJ, Lind MJ, Cawkwell L. The analysis of doxorubicin resistance in human breast cancer cells using antibody microarrays. Molecular cancer therapeutics. 2006;5(8):2115-2120.

[116] Masuda M, Yamada T. Signaling pathway profiling by reverse-phase protein array for personalized cancer medicine. Biochimica et Biophysica Acta (BBA)-Proteins and Proteomics. 2015;1854(6):651-7.

[117] Elemike EE, Uzoh IM, Onwudiwe DC, Babalola OO. The role of nanotechnology in the fortification of plant nutrients and improvement of crop production. Applied Sciences. 2019;9(3):499.

[118] Horton TM, Hoff FW, van Dijk A, Jenkins GN, Morrison D, Bhatla T, et al. The effects of sample handling on proteomics assessed by reverse phase protein arrays (RPPA): Functional proteomic profiling in leukemia. Journal of Proteomics. 2021;233:104046.

[119] Tognon R, Almeida-e-Silva DC, Andraos-Rey R, Ristov M, Ambrósio L, de Almeida FC, et al. A proteomic study of myeloproliferative neoplasms using reverse-phase protein arrays. Leukemia & Lymphoma. 2020;61(13): 3052-3065.

[120] Suzuki M, Muroi A, Nojima M, Numata A, Takasaki H, Sakai R, et al. Utility of a Reverse Phase Protein Array to Evaluate Multiple Biomarkers in Diffuse Large B-Cell Lymphoma. PROTEOMICS–Clinical Applications. 2020;14(1):1900091.

[121] Mahhengam N, fahem ghetran Khazaali A, Aravindhan S, Olegovna Zekiy A, Melnikova L, Siahmansouri H. Applications of Microfluidic Devices in the Diagnosis and Treatment of Cancer: A Review Study. Critical reviews in analytical chemistry. 2021;1-15.

[122] Panesar S, Neethirajan S. Microfluidics: Rapid Diagnosis for Breast Cancer–A Review. Nano-micro letters. 2016;8(3):204-220.

[123] Dai B, Yin C, Wu J, Li W, Zheng L, Lin F, et al. A flux-adaptable pump-free microfluidics-based self-contained platform for multiplex cancer biomarker detection. Lab on a Chip. 2021;21(1): 143-153.

[124] Sancho-Albero M, Sebastián V, Sesé J, Pazo-Cid R, Mendoza G, Arruebo M, et al. Isolation of exosomes from whole blood by a new microfluidic device: proof of concept application in the diagnosis and monitoring of pancreatic cancer. Journal of nanobiotechnology. 2020;18(1):1-15.

[125] Jiang Q, Han T, Ren H, Aziz AUR, Li N, Zhang H, et al. Bladder cancer hunting: A microfluidic paper-based analytical device. Electrophoresis. 2020;41(16-17):1509-1516.

[126] Garcia-Cordero JL, Maerkl SJ. Microfluidic systems for cancer diagnostics. Current opinion in biotechnology. 2020;65:37-44.

[127] Guo Q, Zhang L, Liu J, Li Z, Li J, Zhou W, et al. Multifunctional microfluidic chip for cancer diagnosis and treatment. Nanotheranostics. 2021;5(1):73.

[128] Xu Z, Li E, Guo Z, Yu R, Hao H, Xu Y, et al. Design and construction of a multi-organ microfluidic chip mimicking the in vivo microenvironment of lung cancer metastasis. ACS applied materials & interfaces. 2016;8(39):25840-25847.

[129] Lee H, Cho D-W. One-step fabrication of an organ-on-a-chip with spatial heterogeneity using a 3D bioprinting technology. Lab on a Chip. 2016;16(14):2618-2625.

[130] Yu F, Choudhury D. Microfluidic bioprinting for organ-on-a-chip models. Drug discovery today. 2019;24(6): 1248-1257.

Section 2

Special Cases for Protein Detection

Chapter 4

Multiple and Single Reaction Monitoring Mass Spectrometry for Absolute Quantitation of Proteins

Joshua Yu and Timothy Veenstra

Abstract

The use of mass spectrometry (MS) to measure proteins has grown exponentially over the past 25 years. This growth has been primarily driven by the advent of proteomics in which scientists have developed methods to identify and quantitate as many proteins in a complex mixture as possible. Early studies trended towards the development of techniques that enabled greater quantitative coverage of the proteome. Many of these developments focused on *relative* quantitation in which the change in the abundances of proteins in comparative samples was measured. However, relative quantitation only allows a limited number of samples to be compared. This deficiency led to the development of technologies that allowed the relative quantitation of an unlimited number of samples to be measured, but what was still lacking was an emphasis on the ability of MS to measure the *absolute* abundance of proteins. A more recent technology trend has taken full advantage of the analytical attributes afforded in the use of MS for protein measurements. This trend utilizes the accuracy, sensitivity, specificity, and multiplexed capabilities of MS to quantity specific proteins within complex mixtures. Combined with the use of stable isotope-labeled internal standards, MS assays are now being developed to quantitate key diagnostic and prognostic proteins within clinical samples such as serum, plasma, urine, and cerebrospinal fluid. This chapter describes the technology behind the development of MS-based clinical protein assays and provides examples of where these assays are being used in diagnostic and prognostic settings.

Keywords: mass spectrometry, protein, clinical assay, diagnostics, multiple-reaction monitoring

1. Introduction

Mass spectrometry (MS) has evolved over the past three decades to take its place amongst the premier analytical techniques in use today. The development of MS instrumentation can be traced back to the field of physics, where it was developed as a technique to study the electron [1]. Until the 1940's, MS was still primarily used within the domain of physicists until novel instrument designs by Alfred Nier enabled MS to be used to separate isotopes such as carbon-13 (^{13}C) as well as uranium-235 and uranium-238 [2]. It was during the 1940's that mass spectrometers became commercially available and were widely used by industrial chemists to quantitatively measure molecules whose identities and structures were already known within mixtures [3]. As a result of the efforts of Fred MacLafferty, Klaus Biemann, and Carl Djerassi, the fragmentation mechanisms that occurred during the process of measuring molecules using MS were determined, enabling unknown components within mixtures to be identified [3].

While the analysis of small organic molecules became increasingly routine, it was not until the 1980's that methods for analyzing macromolecules were developed, due to the challenge of finding ways to get large molecules into the gas phase without causing them to fragment. The pioneering work of John Fenn and Koichi Tanaka in the areas of electrospray ionization (ESI) [4] and matrix-assisted laser desorption and ionization (MALDI) [5], respectively, paved the way for macromolecules to be analyzed using MS. It was these developments, especially ESI, from which the field of MS-based proteomics was born.

2. Mass spectrometry and the development of proteomics

The ability to analyze large macromolecules using MS was the beginning of the proteomics revolution. The term proteomics was first coined in 1995 and referred to the identification of individual proteins that had been separated using two-dimensional gel electrophoresis [6]. Initially the proteomics revolution started small, only using MS to identify proteins in simple mixtures [7]. However, as MS technology advanced, the number of proteins identified in increasingly complex biological samples skyrocketed [8–10]. For example, the number of proteins identified in human serum increased from fewer than 500 to over 4000 in less than 20 years [11, 12]. A recent study even identified an astonishing 340,000 proteins across 100 taxonomically diverse organisms, doubling the number of proteins that had previously been identified with direct experiment evidence [13].

As the ability to identify large numbers of proteins became commonplace, advancements in the field turned to focus on quantitation. The initial methods were built to compare the relative abundances of proteins in two samples. The most used method involved stable isotope labeling of proteins that were being translated in cultured prokaryotic or eukaryotic cells [14–16]. In this method, two identical cell lines are cultured *in vitro*. One of the cell lines is cultured in normal media (*e.g.*, DMEM, RPMI, etc.) while the other is cultured in the same medium containing a heavy stable isotope labeled component (*e.g.*, ^{13}C-labeled amino acid, ^{15}N-enriched medium, etc.) [17, 18]. One sample is treated with a specific perturbation, while the other acts as a control. At some time-point the two cultures are combined, the proteins extracted and digested into tryptic peptides, are then analyzed using liquid chromatography (LC) coupled directly online to MS (**Figure 1**). Peptides originating from the separate cultures are distinguished by their heavy isotopic content and are quantitated based on their peak area. This measurement allows the relative abundance of the protein (through its peptide surrogate) in the two samples to be compared and the effect of the perturbation on the protein to be determined. While most of the proteins will not show any change in abundance, studies often identify tens to hundreds of proteins whose abundance changes several-fold due to the perturbation [19, 20].

Once scientists understood that entire proteomes could be analyzed, a variety of methods for conducting comparison studies were quickly developed. Some of

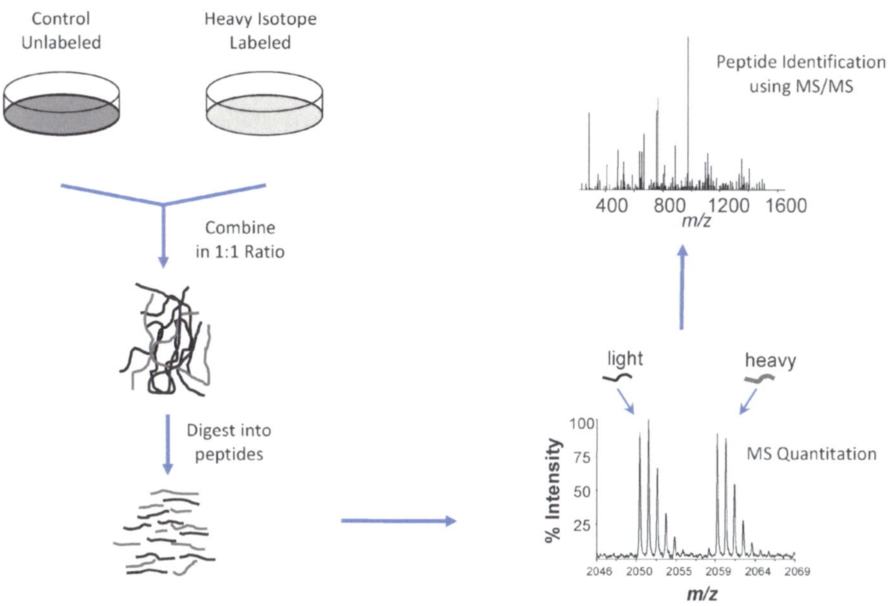

Figure 1.
Use of stable isotope labeling for comparing the relative abundances of proteins in proteome samples. In this method, cells are grown in normal media or media to which a heavy version of a stable isotope has been added. The heavy stable isotope is incorporated into the proteins as they are translated. The proteome samples are combined in equal ratios and the combined sample is proteolytically digested into peptides. The mass spectrum of the isotope labeled and unlabeled peptides are acquired to compare their relative abundance in the two samples. The final step is the automated identification of the peptide using tandem mass spectrometry (MS/MS).

these methods included chemical labeling with isotope tags [21–23], isotope labeling *in vivo* [24], and methods to compare proteomes based on MS peak intensity [25] and number of identified peptides [26]. While label-free methods such as MS peak intensity and peptide number increased the number of samples that could be compared, the samples needed to be prepared and analyzed in a consistent manner for the comparative measurements to be analytically valid. It was not until the development of the Sequential Window Acquisition of all Theoretical Spectra (SWATH) method that the relative abundances of proteins within an unlimited number of samples could be compared [27].

2.1 The biomarker movement

The capability of comparing the relative differences in thousands of proteins in complex samples opened an entire new use for quantitative MS. While much of the focus was on determining how various perturbations changed cellular proteomes, scientists also began to explore how this technology could be used to identify biomarkers of various diseases [28–30]. The premise was simple - acquire a cohort of clinical samples (*e.g.*, plasma, serum, urine, etc.) from a group of patients with a specific disease and a cohort from healthy, matched controls. Each sample would be separately analyzed and the relative abundance of proteins in samples obtained from disease-affected and healthy individuals would be compared [31, 32]. The combined increase in mass spectrometer technology along with this simple sample preparation method resulted in this procedure being utilized extensively in biomarker research. For example, in over the past 50 years, over 30,000 manuscripts containing the keywords "mass spectrometry" and "biomarker" have been listed on PubMed, with almost 97% of these published in the past two decades.

While proteomics has brought about the ability to measure thousands of proteins in hundreds of samples, it also presents several analytical challenges. Since large-scale proteomic biomarker discovery studies can require several weeks, deterioration of overall analytical performance increases in likelihood. For example, chromatographic performance and MS instrument performance can decline as impurities within the biological samples accumulate. A decline in overall performance makes it difficult to determine if differences observed between different cohorts of samples are due to biological or analytical variance. Therefore, robust quality control (QC) strategies are required to optimize the chances of identifying legitimate biomarkers.

Over the past couple of decades there have been thousands of potential biomarkers for a variety of diseases reported in literature. Unfortunately, between the years 1993 and 2008, only 22 novel protein-based tests were approved by the United States Food and Drug Administration (FDA) [33]. There are many issues that have created this discrepancy. Some of them can be understood in the context of how, and how many, samples must be analyzed at each stage of a biomarker discovery project. As shown in **Figure 2**, proteomic biomarker studies generally follow four stages: i) discovery, ii) qualification, iii) verification, and iv) validation. In many ways, the stages are like investigational new drug (IND) clinical trials in scope and number of samples required. The discovery stage typically involves the analysis of tens of samples, however, data on the relative abundance of thousands of analytes is recorded. After this initial data set is scrutinized, only proteins that show a change in abundance between the cohorts are focused on. Any of these proteins that still show a difference between the two cohorts are specifically "targeted" in a larger number of samples. In targeting, the mass

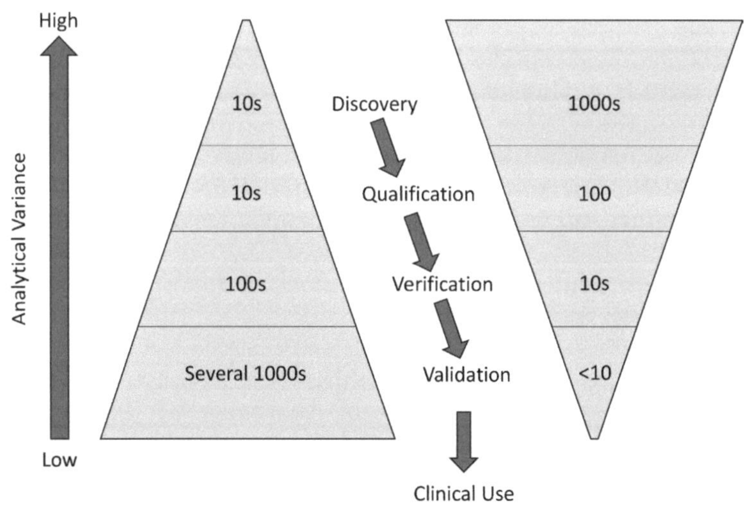

Figure 2.
The steps in biomarker discovery. Biomarker discovery proceeds through the four stages of discovery, qualification, verification, and validation prior to a molecule being reliable for clinical use. Moving through these four stages requires an increasing number of samples to be analyzed, but fewer analytes within each sample to focus on. Not only does the number of samples and analytes vary during these four stages, but so do the analytical and biological variance observed within the samples. The analytical techniques used during the initial discovery stage possess high analytical variance since many analytes must be surveyed. As the study progresses to validation, the analytical variance decreases substantially as the analytical technique chosen is tuned to measure a smaller number of analytes.

spectrometer is instructed to ignore all peaks except those that arise from the qualified proteins that show an abundance difference. Target strategies are aimed at increasing the analytical stringency of measurements, thereby increasing the chance that the abundance difference observed is a result of biological variance. In the final validation stage, specific assays that measure the absolute abundance of a very small number (*i.e.*, less than 10) of proteins are designed and used to analyze thousands of clinical samples. Any protein that passes this final stage may then become approved for clinical use.

One of the biggest hurdles in gaining FDA approval occurs early within the process. The discovery phases of biomarker studies generally result in a significant percentage of the measured proteins showing a difference in their relative abundances between the two cohorts of samples. This feature makes selection of which proteins to focus on in the following stages difficult. Another challenge is one that is common to IND trials: access to proper samples. Progressing through all the stages of a biomarker discovery project ultimately requires access to well over a thousand samples that must be carefully stratified, acquired, stored, and processed. The difficulty in maintaining such a rigorous environment is not a simple task especially when multiple laboratories are involved in the biomarker discovery and validation process. This necessary rigor has been a major contributor to the lack of biomarkers that have been validated using this approach.

2.2 The movement from relative to absolute quantitation

As the number of proteomes and samples that could be compared increased, there was still something lacking in the results: the ability to measure a protein's absolute abundance. While most experiments performed in basic research measure the relative abundance of proteins in samples (*e.g.*, western blotting, immunofluorescence, etc.), knowing the absolute abundance of a molecule allows an unlimited number of samples to be compared. More importantly, measuring the absolute abundance of molecules is the foundation of many medical tests such as comprehensive and basic metabolic panels [34]. If MS were to become a major clinical technique, it also needed to be able to measure the absolute abundance of *proteins*. It was not that the technology was not available: in fact, MS had been used for decades for measuring the absolute quantity of metabolites in clinical samples. Gas chromatography coupled with MS had become mature enough that the U.S. Environmental Protection Agency adopted this technique as its standard method for quantitating several key pollutants [35]. Probably the most well-known quantitative MS assay is the "in-born errors of metabolism" test, which analyze newborn blood samples for defects in fatty acid, organic acid, and amino acid metabolism [36]. While this technique measured metabolites and not proteins, its value is inarguable as millions of infants worldwide and more than 500 confirmed disorders have been screened using this method [36, 37].

3. Absolute protein quantitation

Methods for measuring the quantity of proteins using MS determine either their relative or absolute abundances. While most researchers employ the former approach, there is an increasing movement towards the latter approach. This trend is primarily due to the desire to apply MS for clinical applications. The biggest advantage of measuring a molecule's absolute abundance is the ability to compare results from samples acquired and analyzed anywhere in the world using a standard operating procedure. Absolute quantitation also allows the abundance of different molecules within a single sample to be directly compared. For example, if a cell surface receptor is found to be increased in abundance, a correlation to other proteins involved in its signaling pathway can also be determined if absolute abundances are being measured. Measuring the relative abundance only allows the researcher to determine if a protein's abundance differs between samples taken from different individuals. It does not provide any information related to how a protein's abundance has changed in relation to other proteins within the same sample.

Although antibody-based methods (*i.e.*, immunoassays) currently dominate the protein assay field in clinical laboratories, these methods are not without their disadvantages [38]. Immunoassays can suffer from lot-to-lot antibody variation, high cost, and their need for relatively high sample volumes. In addition, a vast majority of immunoassays only measure a single analyte per experiment. On the positive side, immunoassays are easily automated and there is a large workforce of scientists that are currently trained in conducting these types of experiments.

Using MS to conduct clinical assays has several distinct advantages over immunoassays [39]. Mass spectrometry methods are highly sensitive and require very little sample volume. Since samples are usually fractionated using liquid chromatography prior to MS analysis, hundreds of molecules can be quantitated per experiment. The stable isotope labeled standards, which are added to samples to enable absolute quantitation, are easily synthesized, and do not suffer from lot-to-lot variation.

3.1 Multiple reaction monitoring-mass spectrometry

Determining protein absolute abundance using MS is conducted using a technique called multiple reaction monitoring- or single reaction monitoring-MS (MRM-MS or SRM-MS) [40, 41]. These techniques are sometimes referred to as targeted MS because instead of measuring as many proteins as possible, they are used to quantify specific (or targeted) proteins in a mixture. While enzyme-linked immunosorbent assays (ELISAs) have been the dominant technique for measuring the absolute abundance of protein in complex mixtures, MRM- and SRM-MS methods offer several key advantages. MRM- and SRM-MS assays are not limited by the availability of suitable antibodies. This advantage allows these assays to be readily customizable to suit any new targets that are discovered. MRM- and SRM-MS assays also provide a direct measurement of the analyte of interest, whereas ELISAs can suffer from cross-reactivity associated with antibodies [42].

MRM-MS is generally performed using triple quadrupole or triple quadrupole ion trap mass spectrometers because of their linear quantitation range, however, other types of instruments (*e.g.*, Orbitraps, quadruple-TOFs, etc.) can also be used [43, 44]. As shown in **Figure 3**, the first step of an MRM-MS (and SRM-MS) experiment is to digest the proteome into peptides [43–45]. The peptides are then separated using LC coupled directly on-line with the mass spectrometer. As they elute into

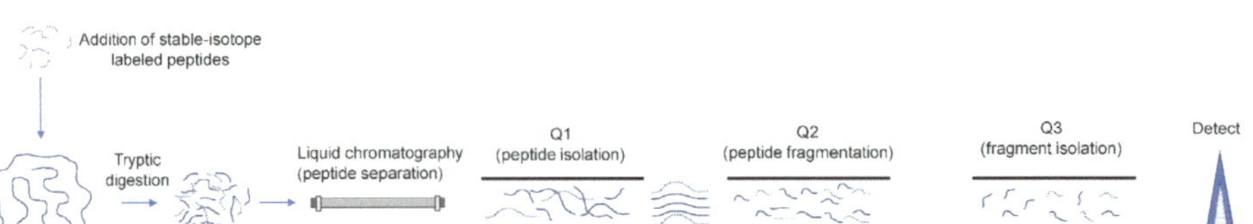

Figure 3.
Multiple reaction monitoring-mass spectrometry (MRM-MS) for the absolute quantitation of proteins.

the first quadrupole (Q1), the specific peptide of interest is isolated by applying a radiofrequency/direct current (RF/DC) potential that prevents all other peptides from passing through this region. The isolated peptide then moves onto the second quadrupole (Q2) and is fragmented, generally through collisions with an inert gas. All the fragments enter the third quadrupole (Q3), however, only around 3–5 are isolated and detected. A SRM-MS experiment is done in a very similar manner, except only a single fragment is isolated in Q3 and detected.

So why are 3–5 fragments selected in an MRM-MS experiment but only one in a SRM study? The answer partly comes down to accuracy. Ensuring that the correct peptide is being measured is foundational to quantitative proteomics. It is inadequate to identify a peptide only based on molecular weight (MW) because the complexity of a proteome sample means many peptides will have the same (or very similar) MWs. It is unlikely, however, that two peptides will have identical fragment ions. In MRM-MS, multiple fragment ions are monitored to ensure the correct peptide is being measured, while in SRM-MS, only a single fragment ion is being monitored. In general MRM-MS is used regularly for peptide measurements while SRM-MS is used for measuring small metabolites. Therefore, the rest of this chapter will focus only on MRM-MS quantitative studies.

To measure the absolute quantity of the peptide, a known amount of stable isotope labeled internal standard with the same sequence as the peptide of interest is added to the sample. The amount of endogenous peptide can then be determined by comparing its peak area to that of the internal standard.

As with any assay, an MRM-MS assay is only useful if it is specifically measuring the protein of interest. Therefore, rules have been established to ensure the specificity of the MRM-MS assay [46]. The peptide selected to act as the surrogate must be unique to the targeted protein. To ensure its uniqueness, the sequence of the peptide must be compared to all sequences within the corresponding species-specific protein database. The peptide should be between 7 and 20 amino acids long since this size range is ideal for efficient generation of fragments. It should also be within all the different isoforms of the protein (if any). If the experiment is designed to measure the abundances of different isoforms of a protein, the peptides selected need to be unique to each isoform. To prevent any change to the peptide during sample processing, it should be void of any easily oxidizable amino acids (*i.e.*, cysteine, methionine, etc.). In addition, it is best to measure multiple (*i.e.*, 3–5) peptides from the same protein whenever possible so that the results from each can be used to confirm the result.

3.2 Validating MRM-MS

Prior to its widespread use, MRM-MS needed to prove itself as an accurate technique for measuring protein abundance. As the goal of using MRM-MS was to measure biomarkers in clinical samples, the robustness of the technique had to be extensively validated in serum, plasma, and urine samples [47–49]. In one of these validation studies [47], eight participating laboratories were asked to measure the absolute abundance of 11 peptides originating from 7 different proteins. The aim of this study was to determine the intra- and inter-laboratory variation of MRM-MS assays at different sites. Five of the proteins were non-human in origin, to eliminate any unpredictable interference from endogenous proteins already within the plasma sample. The other two proteins were the prostate cancer biomarker, human prostate-specific antigen (PSA) [50] and the acute phase response protein, C-reactive protein (CRP) [51].

This study was conducted in three phases [47], with the measured outcome of each phase being the absolute recovery of the peptides based on comparison to the signals generated by the heavy isotope labeled internal standards. The three phases differed in the amount of sample preparation required at each site. In phase I, the participating labs received plasma samples pre-digested using trypsin that already contained the 11 peptides of interest. Each lab had to simply run the MRM-MS assay. The inter- and intra-laboratory reproducibility and precision in this phase was excellent for all peptides. In phase II, each laboratory had to perform the tryptic digestion step and clean up the sample prior to MRM-MS analysis. The additional sample preparation steps had minimal impact on the precision as the coefficient of variation (CV) between the labs remained less than 15%. In phase III, all the sample preparation step were performed on-site (*i.e.*, additional of heavy isotope labeled internal standards, tryptic digestion, sample clean-up, etc.) prior to MRM-MS analysis. While the overall peptide recovery was significantly lower (*i.e.*, 119.8%, 79.6%, and 48.9% for phases I, II, and III, respectively), the interlaboratory CVs remained below 25% for 8 of the 11 peptides of interest. This study highlighted areas of special concern, primarily in sample preparation steps, that need to be standardized to ensure accuracy in absolute protein abundance measurements.

3.3 Multiplexing the technology

The combination of LC separation with MS analysis provides an analytical platform capable of detecting thousands of individual components in a clinical sample such as plasma, serum, or urine. Imagine what could be done if many of these components were valid biomarkers for various conditions such as cancer and even a fraction of these detectable components could be accurately quantitated. Patients could provide blood and urine samples at yearly physicals that could be used to diagnose specific cancer types at very early stages. As early detection is a key to surviving cancer, the ability to accurately quantify specific biomarkers in routinely acquired clinical sample would substantially decrease the death rate due to many different cancers.

While this multiplexing capability is currently employed for measuring many inborn errors such as metabolism [36], steroid hormones [52], and lipids [53], proteins are not included. However, there is a current effort to expand its use to proteins, and in the largest study to date, an assay to quantitatively measure 267 proteins using MRM-MS was designed [54]. To test its veracity, the 267 proteins were measured across 21 different commercially available human plasma samples. Within these proteins, 61 had been FDA-approved for use in laboratory developed tests (LDTs) [55] and a further 67 were putative cardiovascular disease biomarkers [56]. The major goal of this study, beyond showing the utility of MRM-MS as an analytical tool, was to develop an easy-to-use standardized kit that would provide reproducible and transferable clinical results in a variety of laboratory settings.

In this study, a known amount of stable isotope labeled peptides corresponding to specific endogenous target peptides was added to each commercial plasma sample. The absolute abundances were then calculated by determining the peak area ratios between the endogenous target peptides and their corresponding stable isotope standard and comparing this ratio to previously constructed standard curves. The standard curves were prepared on all three days using 5–8 standards. These curves allowed both the lower and upper limits of quantitation (LLOQ and ULOQ) to be calculated for each protein. Each sample was analyzed each day for three separate days to evaluate robustness. Of the 267 targeted proteins, 144 were quantified in at least 5 of the 21 samples. Just over one-third of these were FDA approved as biomarkers or for use in LDTs. In addition, a total of 111 proteins were quantified in all 21 samples over the course of the three-day analysis stage. No protein in any of the samples had a concentration above the ULOQ, however, 110 proteins were below the LLOQ in all 21 samples in all 3 separate analyses. Over half of the proteins whose concentrations were below the LLOQ have associations with various conditions including cardiomyopathy, epilepsy, neuropathy, rheumatoid arthritis, and lung cancer. Although a measurement below the LLOQ prevents a protein from being quantified, this observation is still important for evaluating various diseases in which a specific protein may be upregulated, and its concentration falls within the limit of detection (LOD) of this multiplexed assay.

As the goal of this study was to develop a kit that could be used across various laboratories, it was important that the assay possessed consistency. Comparison of the 21 samples showed high consistency as a median of 145 proteins (standard deviation 15.8) were quantified over the three days of analysis, as shown in **Figure 4A.**, with an 84% overlap of quantifiable proteins per sample over the three analyses. Overall, the data showed that the MRM-MS assay is highly reproducible for quantifying multiple protein targets in a very complex biological sample. The CVs were calculated for all the proteins that could be quantified (**Figure 4B**), and approximately 70% of the

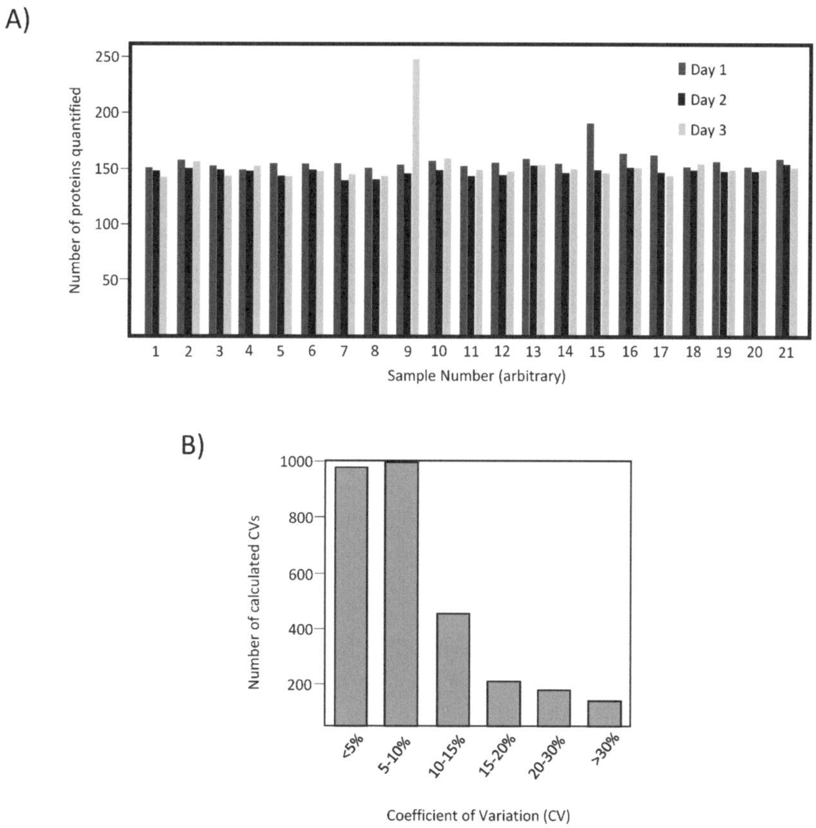

Figure 4.
(A) Number of proteins that were quantified within 21 plasma samples analyzed over a three-day period. To be counted, each protein had to meet acceptable analytical criteria. (B) Distribution of percentage of quantitated proteins based on their coefficients of variation (CVs) of their abundances measured over three days.

CVs were below 10%, with only about 8% greater than 20%. Taken together, the data points towards the reliability of multiplexed MRM-MS assays quantitating proteins within clinical laboratories.

Another important characteristic of any analytical technique, specifically one aiming to be used to measure biomarkers, is its dynamic range. Dynamic range refers to the range of concentrations that a technique can measure [57]. Since the concentration range of proteins in human plasma spans 12 orders of magnitude, it is important that MRM-MS provide a comparable dynamic range [58]. The developed MRM-MS assay displayed a dynamic range of six orders of magnitude ranging from a high of serum albumin (747 pmol) to a low of p-selectin (1 fmol). P-selectin is up-regulated in endothelial cells and platelets in sickle-cell anemia patients and contributes to their symptoms of pain and vaso-occlusion [59]. Recently, Novartis was granted approval for the use of crizanlizumab as a treatment for reducing these symptoms through its binding to p-selectin [60, 61]. The coefficients of variation for 86% of the measured proteins were less than 15%. The differences in protein concentrations in the 21 samples ranged from a highly consistent 1.1-fold for metalloproteinase inhibitor 2 to a widely variable 69-fold for serum amyloid A1/A2. The highest variability observed across the samples for an FDA-approved biomarker was the 60-fold variation seen for C-reactive protein, which along with serum amyloid A is known to increase several hundred-fold in concentration due to an inflammatory response [62].

Working with MRM Proteomics, Inc. and Cambridge Isotopes Laboratories, Inc., investigators coupled the necessary reagents and instructions into the commercially available PeptiQuant™ Assay Kits for perform the analysis described above [63]. The kits are available as either PeptiQuant Plus Quality Control Kits for evaluating the performance of a lab-developed MRM-MS assay or the PeptiQuant Plus Biomarkers Assessment Kit that analyzes human or mouse plasma samples for more than 125 disease-related biomarkers. The development of these kits not only shows that MRM-MS is an important tool for diagnosing and monitoring diseases such as cancer, heart disease, etc. but the commercialization of the PeptiQuant Assay Kits demonstrates that the actual clinical use is not far off in the future.

4. Applications of MRM-MS

An immediate use of MRM-MS is the improvement on existing techniques for measuring known disease biomarkers. Hepatocyte growth factor receptor (HGFR), also known as Met, is a tyrosine kinase membrane receptor that is measured as a biomarker for various cancers [64]. When over-activated, Met activates several biological activities that can result in an invasive oncogenic phenotype. The overexpression of Met is directly correlated with tumor aggression and poor patient outcome. Such correlation has been observed in gastroesophageal and esophageal adenocarcinomas [65]. As a result, Met is a prime therapeutic target and several monoclonal antibodies (mAb) that bind to and inhibit either HGF or Met are currently being tested in clinical trials. In fact, a patient with stage IV gastroesophageal cancer that had a high *MET* gene copy number and concordant Met expression showed complete response to the Met-specific mAb onartuzumab [66]. In a phase II trial, rilotumumab treatment increased the survival of patients with tumors exhibiting a high Met expression level compared to those with low Met levels [67].

Selection of the correct treatment of Met-driven tumors depends on the ability to measure Met expression in patient samples. These measurements are routinely performed using immunohistochemistry (IHC) [68]. Unfortunately, IHC is only semi-quantitative; simply providing a "score" of 0, +1, +2, or + 3 that signifies the levels of protein expression. The score is generated by considering the staining intensity as well as the percentage of stained cells. The score is also prone to subjective bias since the colorimetric signal produced during IHC is interpreted by a pathologist [69]. While imaging software has attempted to standardize the technique, IHC is still not uniformly and systematically applied for quantitating specific proteins in tumor biopsies.

To overcome the subjectivity and semi-quantitative nature of IHC, Liquid-Tissue-SRM was developed specifically to measure the absolute abundance of specific proteins in formalin-fixed paraffin-embedded (FFPE) tissue [70–72]. In this method, tumor cells are extracted from tissue sections using laser microdissection (**Figure 5**). A lysate is prepared from these cells and digested into peptides using trypsin, and a stable-isotope labeled version of a peptide within the target protein (in this case Met) is added to the peptide mixture and the sample is analyzed using LC-SRM-MS.

To test the method, 130 FFPE gastroesophageal cancer tissues were analyzed using Liquid Tissue, IHC, mean *MET* gene copy number/nucleus, and *MET/CEP7* gene copy number ratio using fluorescence *in situ* hybridization (FISH) [72]. The correlation between Liquid Tissue and IHC results was low (R^2 = 0.537). The correlation between Liquid Tissue results and the *MET*/nucleus and *MET/CEP7* gene copy number results, however, was high (R^2 = 0.898).

In a similar study, an MRM-MS method was developed to quantitate human epidermal growth factor receptor 2 (HER2) in breast cancer tumors [73]. Overexpression of HER2 is associated with breast cancer and higher levels of the protein correlate with poorer patient outcomes [74]. While the study ultimately sought to develop a method with higher quantitative accuracy than IHC, it also was attempting to address a limitation with FISH. Specifically, FISH is routinely used to evaluate samples assigned an IHC score of +2. Unfortunately, the process is time-consuming and expensive. Therefore, establishing an MRM-MS method that obviated the need for both IHC and FISH would improve on existing methods for differentiating HER2 status as well as reducing costs.

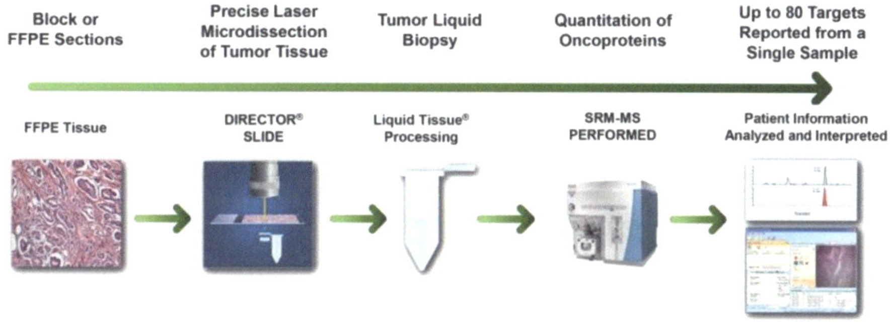

Figure 5.
Liquid tissue-selected reaction monitoring-mass spectrometry (SRM-MS) workflow for quantitation of proteins extracted from formalin-fixed paraffin-embedded (FFPE) tissue sections. In this method, specific areas procured from deparaffinized tissue sections are placed in a tube. The proteins within the cells are extracted and digested into peptides. A known amount of heavy isotope labeled peptide corresponding to a peptide of interest is added and the samples are analyzed using SRM-MS to measure the absolute abundance of the targeted endogenous peptide.

In this study, 210 FFPE tissue sections were cut from breast cancer samples [73]. After processing the samples using a similar Liquid Tissue process as described above, the absolute levels of HER2 and the epithelial cell-specific protein, adhesion molecule A, were measured using MRM-MS. To normalize the results, the absolute amount of HER2 was divided by that of adhesion molecule A, which is not expected to change based on tumor status. In contrast to IHC alone, the MRM-MS analysis was able to distinguish HER2 2+/FISH positive and HER2 2+/FISH negative samples, with an area under the receiver operator characteristic curve (AUROC) of 0.908. The results show that MRM-MS assays provide more accurate HER2 expression levels than IHC and can eliminate uncertainty in the decision making of oncologists seeking the proper treatment of HER2 positive breast tumors.

4.1 Taking MRM *in vivo*

While measuring the absolute abundance of a protein is a hallmark of diagnosing and/or monitoring many disease states, there is value in measuring protein turnover as well, since it can detect abnormalities in protein clearance rates and anticipate potential build up of proteins in the future. While protein turnover could be measured by adding stable isotope labeled standards to a series of samples acquired at several time points, this method suffers from analytical variabilities and places a significant burden on the patient. To overcome these challenges, investigators at Washington University in St. Louis developed an *in vivo* stable isotope labeling method to study protein turnover. While stable isotope labeling *in vivo* had been done previously on animal species including rats and mice [75], this was the first case of the procedure being performed on humans [76]. This method, termed stable isotope labeling kinetics (SILK), used *in vivo* incorporation of a stable isotope labeled amino acid to measure the fractional synthesis (FSR) and fractional clearance rates (FCR) of specific proteins. The focus of SILK was to measure the FSR and FCR of beta-amyloid (Aβ) to evaluate the role that this protein's production and clearance plays in plaque formation [77].

In the SILK procedure, human subjects are intravenously infused over a 9-hour period with $^{13}C_6$-labeled leucine ($^{13}C_6$-Leu) dissolved in normal saline (**Figure 6**). As $^{13}C_6$-Leu enters the blood stream, it is incorporated into proteins that are being actively translated within cells. Blood samples are then taken from the patient every hour for up to 48 hours. This time provides the data necessary to measure the uptake of $^{13}C_6$-Leu into newly translated Aβ during the earlier time points and the degradation of this protein population over the later time points. To simplify the mixture, an antibody is used to immunoprecipitate Aβ from each sample. The purified Aβ is then digested into peptides and analyzed using MRM-MS.

The MRM-MS data yields the quantitative ratio between the $^{13}C_6$-Leu labeled and unlabeled Aβ peptides. During the early time points (*i.e.*, between about hours 8–16), this ratio increases as the pool of $^{13}C_6$-Leu in the cell increases [77]. Measuring the slope of this increase allows the FSR to be calculated. Later time points see a drop in the $^{13}C_6$-labeled/unlabeled ratio as the pool of $^{13}C_6$-Leu is diminished (*i.e.*, between about hours 20–28), and proteins that had incorporated the heavy amino acid start to be degraded. The FCR is calculated by measuring the slope of this decreasing ratio. Any changes in the FSR or FCR could potentially lead to a diagnosis of Alzheimer's disease, a notoriously difficult condition to diagnose in early stages - when intervention would have the greatest impact in preventing its progression.

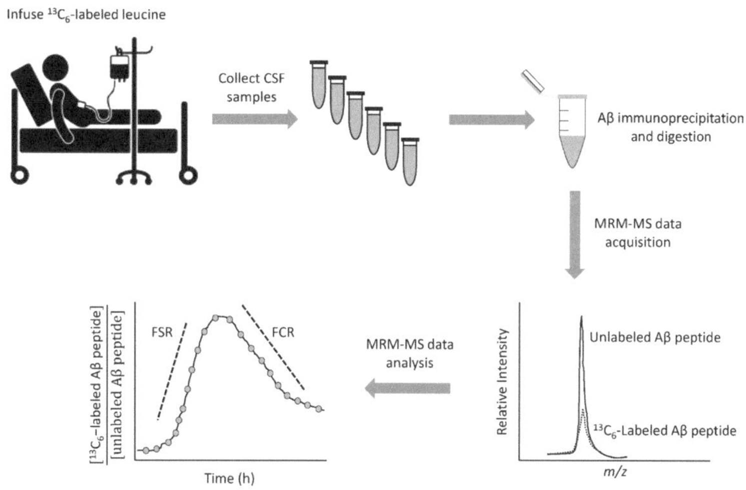

Figure 6.
Methodology for conducting stable isotope labeling kinetics (SILK) study to determine fractional synthesis (FSR) and fractional clearance rates (FCR) of beta-amyloid (Aβ) in vivo. In a SILK study, a solution of $^{13}C_6$-labeled leucine ($^{13}C_6$-Leu) solution is given intravenously to the subject. Aliquots of cerebrospinal fluid (CSF) are acquired at various timepoints after initiation of the $^{13}C_6$-Leu infusion. Amyloid β is immunoprecipitated from the CSF samples and digested into smaller peptides prior to multiple reaction monitoring mass spectrometry (MRM-MS) analysis. The ratio of the $^{13}C_6$-labeled peptide to its unlabeled counterpart is plotted to calculate the FSR and FCR of Aβ in the subject.

4.2 Microbial identification

Bacteremia, a bacterial infection in blood, is a serious condition as it allows the bacteria to spread to any part of the body [78]. The infection can lead to sepsis, a dysregulated response to an infection, and remains a major cause of hospital morbidity and mortality [79]. The conventional method of identifying bloodstream infections require overnight subculturing followed by identification of the relevant species and antimicrobial susceptibility testing. Although automated, this process still requires about 18–24 hours to complete [80]. The MALDI time-of-flight (MALDI-TOF) MS procedure can provide identification results in less than an hour using bacterial pellets obtained from blood culture vials. This reduction in time can have a tremendous impact on survival of patients with sepsis. Since there is an inverse relationship between the time to sepsis diagnosis and patient mortality, decreasing the time required to identify bacteria-specific infections would be expected to have a major beneficial impact on public health [81–83].

Arguably the most successful application of MS in the clinic is the use of MALDI-TOF MS for the identification of microorganisms. While the identification of bacteria using MS dates to the early 1970's, the techniques then relied mainly on the detection of lipids of microorganisms grown in agar [84]. It was in the 1990's that the ability to obtain MALDI-TOF MS profiles of the cell contents of bacteria after sonication was demonstrated [85]. This ability led to the generation of "fingerprints" that could be assigned to specific bacterial types, which enabled rapid identification of bacteria using MS [86]. MALDI-TOF MS methods are becoming increasingly commonplace in clinical microbiology laboratories, because of its capability to reduce identification times by over 75% [87].

Sample preparation in MALDI-TOF MS procedures is simple [88]. Briefly, 1 mL of blood culture broth is mixed and incubated for 2 minutes with an extraction strip

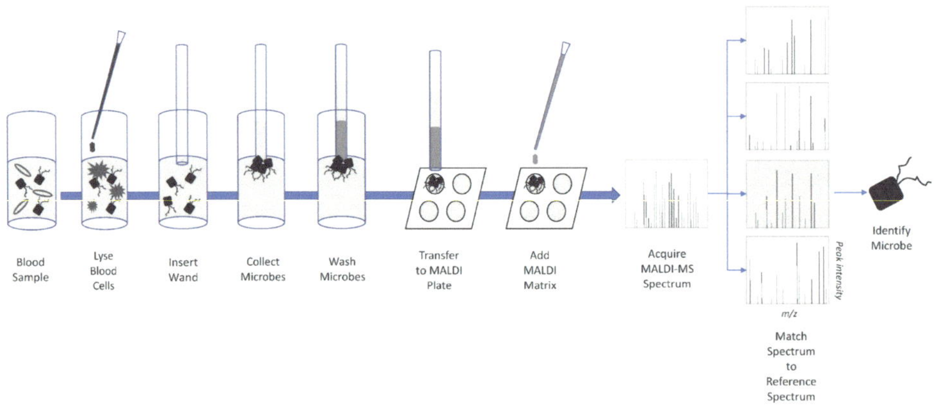

Figure 7.
Determination of microbes in blood using MALDI-MS. A buffer is added to blood to lyse red blood cells followed by the insertion of a filter wand. Application of a vacuum to the wand draws the microbes to its surface, where they are washed prior to being spotting on a target of a MALDI plate. After addition of the MALDI matrix, the mass spectrum of the microbes is recorded. To identify the unknown microbe, the recorded spectrum is compared to a reference library of spectra acquired of known microbes.

that contains a lysis buffer (**Figure 7**). This treatment lysis red blood cells but allows microorganisms to remain intact. The microorganisms are then captured onto a membrane surface by immersing a filter wand into the lysate and applying a vacuum. After washing using a series of buffers, the microorganisms attached to the filter are transferred to a MALDI plate target spot by by tapping the filter onto the surface of the plate. Cyano-4-hydroxycinnamic acid and formic acid are then applied to the spots containing the microbial sample, which is allowed to dry prior to MALDI-TOF MS data acquisition.

After acquisition of the MALDI-TOF MS spectra, microorganisms contained within the sample are identified by matching the acquired protein and peptide peak profiles to reference profiles contained within annotated databases. A previous study showed that the proteins detected using this method are primarily composed of ribosomal, DNA-binding, and cold shock proteins [89].

5. Conclusions

The past 30 years has seen a major shift in protein science. Before 1990, protein science was almost exclusively a "one-at-a-time" process in which a feature (*e.g.*, abundance, size, structure, binding partner, etc.) of a single protein was studied per experiment. With the advent of ESI and the coupling of LC with MS, the relative abundance of thousands of proteins could now be compared in complex biological samples. This capability accelerated the growth of proteomics and resulted in several investigators using this technology to search for biomarkers of various diseases. Unfortunately, as the data accumulated it became evident that potential biomarkers could only be validated if their absolute abundance could be measured.

The obvious competitor to MRM-MS assays are antibody-based methods such as ELISAs. ELISAs are routinely performed and with the use of standards, can provide results that measure the absolute abundance of a specific protein. However, many scientists argue that MRM-MS methods are superior owing to their specificity. Not only can antibody measurements suffer from cross-reactivity, but there are not antibodies available for every protein isoform currently known. While there are some protein isoforms that MRM-MS may not be able to measure, such as specific phosphorylated forms or those with very high homology, it is generally conceded that MRM-MS assays can be developed for a broader range of proteins and other types of biomolecules.

While MRM-MS does have several advantages over antibody-based techniques, it is not without its disadvantages. MRM-MS assays may not always have the sensitivity to match the endogenous level of the target protein within its biological setting. In addition, the signal produced by MS is often negatively impacted by the presence of other components in the mixture. Therefore, extensive prefractionation or molecule-specific enrichment of serum, plasma, and tissue samples may be necessary to optimize the LLOQ of an MRM-MS assay. If the goal is to develop an MRM-MS assay that measures a panel of different proteins, the fractionation and/or enrichment steps need to be applicable to all the targeted species. Incorporation of any additional steps not only decreases the throughput but introduces additional analytical variability. Since antibody-based methods generally do not require additional prefractionation or enrichment steps, their throughput is more constant.

Another disadvantage of MRM-MS methods is cost. As MRM-MS capabilities increased with the development of more sensitive, higher resolution mass spectrometers, the cost of these instruments likewise increased. It is not uncommon for a typical LC–MS system necessary for conducting MRM-MS assays to cost more than 400,000 U.S. dollars. Along with the costs of reagents (especially stable isotope standards) and software, MRM-MS represents a significant cost per analysis. While MRM-MS has multiplexing capabilities, only one sample can be run per instrument, whereas hundreds (or thousands) of samples can be analyzed on a single ELISA plate. Increasing sample throughput for an MRM-MS assay requires purchasing additional LC/MS systems, which is an additional major capital equipment expense.

MRM-MS methods for measure the absolute quantity of proteins will continue their rapid progress and likely become routinely used in diagnostic and prognostic medicine. Their ability to measure multiple physiological biomarkers in easily acquired biological samples represents a tremendous advancement in diagnostic medicine. Imagine giving a blood sample at an annual physical and having it analyzed for biomarkers that are diagnostic for various cancers, myopathies, cardiomyopathies, neurological disorders, inflammation, etc. Routine detection of these conditions at an early stage would have an enormous impact on life expectancy. This scenario is the ultimate advantage of MRM-MS assays over antibody-based tested in future medicine; no hypothesis would be required to identify a current disease condition. Much like how genome-wide sequencing is becoming highly accessible to the public, quantitative protein screening via MRM-MS assays will soon follow.

Acknowledgements

The authors would like to thank Cedarville University for support during the preparation of this chapter.

Conflict of interest

The authors declare no conflict of interest.

Author details

Joshua Yu[1] and Timothy Veenstra[2*]

1 Phillips Exeter Academy, Exeter, NH, USA

2 School of Pharmacy, Cedarville University, Cedarville, OH, USA

*Address all correspondence to: tveenstra@cedarville.edu

References

[1] Dahl PF. Flash of the Cathode Rays: A History of J. J. Thomson's Electron; Institute of Physics Publishing: Bristol, U.K., 1997. DOI:10.1063/1.882210

[2] Nier AO. Some reminiscences of mass spectrometry and the Manhattan project. J. Chem. Educ 1989;66:385-388. DOI:10.1021/ed066p385

[3] Reinhardt, C. Shifting and Rearranging: Physical Methods and the Transformation of Modern Chemistry; Science History Publications: Sagamore Beach, MA, USA, 2006.

[4] Fenn JB, Mann M, Meng CK, Wong SF, Whitehouse CM. Electrospray ionization for mass spectrometry of large biomolecules. Science 1989;246:64-71. DOI:10.1126/science.2675315

[5] Tanaka K, Waki H, Ido Y, Akita S, Yoshida Y, Yoshida T. Protein and polymer analyses up to *m/z* 100 000 by laser ionization time-of-flight mass spectrometry. Rapid Commun. Mass Spectrom 1988;2:151-153. DOI:10.1002/rcm.1290020802

[6] Wasinger VC, Cordwell SJ, Cerpa-Poljak A, Yan JX, Gooley AA, Wilkens MR, Duncan MW, Harris R, Williams KL, Humphery-Smith I. Progress with gene-product mapping of the Mollicutes: Mycoplasma genitalium. Electrophoresis 1995;16:1090-1094.

[7] Wang R, Chait BT. High-accuracy mass measurement as a tool for studying proteins. Curr, Opin. Biotechnol 5;1994:77-84. DOI: 10.1016/s0958-1669(05)80074-6

[8] Washburn MP, Wolters D, Yates JR 3rd. Large-scale analysis of the yeast proteome by multidimensional protein identification technology. Nat. Biotechnol 2001;19:242-247. DOI: 10.1038/85686

[9] Francischetti IMB, Mans BJ, Meng Z, Gudderra N, Veenstra TD, Pham VM, Ribeiro JMC. Insect Biochem. Mol. Biol. 2008;38:1-21. DOI: 10.1016/j.ibmb.2007.09.009

[10] Walther TC, Mann M. Mass spectrometry-based proteomics in cell biology. J. Cell Biol 2010;190:491-500. DOI: 10.1083/jcb.201004052

[11] Tirumalai RS, Chan KC, Prieto DA, Issaq HJ, Conrads TP, Veenstra TD. Characterization of the low molecular weight human serum proteome. Mol. Cell. Proteomics 2003;2:1096-1103. DOI: 10.1074/mcp.M300031-MCP200

[12] Uyar DS, Huang YW, Chesnik MA, Doan NB, Mirza SP. Comprehensive serum proteomic analysis in early endometrial cancer. J. Proteomics 2021;234:104099-104107. DOI: 10.1016/j.jprot.2020.104099r.

[13] Müller JB, Geyer PE, Colaço AR, Treit PV, Strauss MT, Oroshi M, Doll S, Virreira Winter S, Bader JM, Kohler N, Theis F, Santos A, Mann M. The proteome landscape of the kingdoms of life. Nature 2020;582:592-596. DOI: 10.1038/s41586-020-2402-x

[14] Martinovic S, Veenstra TD, Anderson GA, Paša-Tolic L, Smith RD. Selective incorporation of isotopically labeled amino acid for identification of intact proteins on a proteome-wide level. J. Mass Spectrom 2002;37:99-107.

[15] Ong SE, Blagoev B, Kratchmarova I, Kristensen DB, Steen H, Pandey A, Mann M. Stable isotope labeling by

amino acids in cell culture, SILAC, as a simple and accurate approach to expression proteomics. Mol. Cell. Proteomics 2002;1:376-386. DOI: 10.1074/mcp.m20025-mcp200

[16] Aebersold R, Mann M. Mass spectrometry-based proteomics. Nature 2003;422;198-207. DOI: 10.1038/nature01511

[17] Rozanova S, Barkovits K, Nikolov M, Schmidt C, Urlaub H, Marcus K. Quantitative mass spectrometry-based proteomics: an overview. Methods Mol. Biol 2021;2228:85-116. DOI: 10.1007/978-1-0716-1024-4_8

[18] Otto A. Metabolic labeling of microorganisms with stable heavy nitrogen isotopes (15N). Methods Mol. Biol 2018;1841:175-188. DOI: 10.1007/978-1-4939-8695-8_13

[19] Zanivan S, Maione F, Hein MY, Hernandez-Fernaud JR, Ostasiewicz P, Giraudo E, Mann M. SILAC-based proteomics of human primary endothelial cell morphogenesis unveils tumor angiogenic markers. Mol. Cell. Proteomics 2013;12:3599-3611. DOI: 10.1074/mcp.M113.031344

[20] Hart JR, Zhang Y, Liao L, Ueno L, Du L, Jonkers M, Yates JR 3rd, Vogt PK. The butterfly effect in cancer: a single base mutation can remodel the cell. Proc. Natl. Acad. Sci. U.S.A 2015;112:1131-1136. DOI: 10.1073/pnas.1424012112

[21] Gygi SP, Rist B, Gerber SA, Turecek F, Gelb MH, Aebersold RA. Quantitative analysis of complex protein mixtures using isotope-coded affinity tags. Nat. Biotechnol 1999;17;994-999. DOI: 10.1038/13690

[22] Goshe MB, Veenstra TD, Panisko EA, Conrads TP, Angell NH, Smith RD. Phosphoprotein isotope-coded affinity-tags: application to the enrichment and identification of low-abundance phosphoproteins. Anal. Chem 2002;74:607-616. DOI: 10.1021/ac15528g

[23] Li K, Chen Z, Zhang Y, Yang X. Applications of iTRAQ and TMT labeling techniques to the study of neurodegenerative diseases. Curr. Protein Pept. Sci 2020;21:1202-1217. DOI: 10.2174/1389203721666201103085704

[24] Liao L, McClatchy DB, Park SK, Xu T, Lu B, Yates JR 3rd. Quantitative analysis of brain nuclear phosphoproteins identifies developmentally regulated phosphorylation events. J. Proteome Res 2008;7:4743-4755. DOI: 10.1021/pr8003198

[25] Ankney JA, Muneer A, Chen X. Relative and absolute quantitation in mass spectrometry-based proteomics. Annu. Rev. Anal. Chem 2018;11:49-77. DOI: 10.1146/annurev-anchem-061516-045357

[26] Schirmer EC, Florens L, Guan T, Yates JR 3rd, Gerace L. Nuclear membrane proteins with potential disease links found by subtractive proteomics. Science 2003;30:1380-1382. DOI: 10.1126/science.1088176

[27] Gillet LC, Navarro P, Tate S, Rost H, Selevsek N, Reiter L, Bonner R, Aebersold R. Targeted data extraction of the MS/MS spectra generated by data-independent acquisition: a new concept for consistent and accurate proteome analysis. Mol. Cell. Proteomics 2012;11:O111.016717. DOI: 10.1074/mcp.O111.016717

[28] Veenstra TD. Global and targeted quantitative proteomics for biomarker discovery. J. Chromatogr. B. Analyt. Technol. Biomed. Life Sci 2007;847:3-11. DOI: 10.1016/j/jchromb.2006.09.004

[29] Jimenez CR, Piersma S, Pham TV. High-throughput and targeted in-depth

mass spectrometry-based approaches for biofluid profiling and biomarker discovery. Biomarker Med 2007;1:541-565. DOI: 10.2217/17520363.1.4.541

[30] Dagley LF, Emili A, Purcell AW. Application of quantitative proteomics technology to the biomarker discovery pipeline for multiple sclerosis. Proteomics Clin. Appl 2013;7:91-108. DOI: 10.1002/prca.201200104

[31] Wang M, You J. Mass spectrometry for protein quantification in biomarker discovery. Methods Mol. Biol 2012;815:199-225. DOI: 10.1007/978-1-61779-424-7_16

[32] Shao S, Guo T, Aebersold RA. Mass spectrometry-based proteomic quest for diabetes biomarkers. Biochim. Biophys. Acta 2015;1854:519-527. DOI: 10.1016/j.bbapap.2014.12.012

[33] Parker CE, Borchers CH. Mass spectrometry-based biomarker discovery, verification, and validation – quality assurance and control of protein biomarker assays. Mol. Oncol 8;2014:840-858. DOI: 10.1016/j.molonc.2014.03.006

[34] Prezioso D, Di Martino M, Galasso R, Iapicca G. Laboratory assessment. Urol. Int 2007;1:20-25. DOI: 10.1159/000104437

[35] Budde WL. Ron Hites: gas chromatography/mass spectrometry pioneer and the great detector debate. Environ. Sci. Technol 2015;49:13741-13742. DOI: 10.1021/es505993z

[36] Naylor EW, Chace DH. Automated tandem mass spectrometry for mass newborn screening for disorders in fatty acid, organic acid, and amino acid metabolism. J. Child. Neurol 1999;14 Suppl 1:S4-S8. DOI: 10.1177/0883073899014001021

[37] Chace DH, Kalas TA, Naylor EW. The application of tandem mass spectrometry to neonatal screening for inherited disorders of intermediary metabolism. Annu. Rev. Genomics Hum. Genet 2002;3:17-45. DOI: 10.1146/annurev.genom.3.022502.103213

[38] Hoofnagle AN, Wener MH. The fundamental flaws of immunoassays and potential solutions using tandem mass spectrometry. J. Immunol. Methods 2009;347:3-11. DOI: 10.1016/j.jim.2009.06.0043

[39] Percy AJ, Chambers AG, Parker CE, Borchers CH. Absolute quantitation of proteins in human blood by multiplexed multiple reaction monitoring mass spectrometry. Methods Mol. Biol 2013;1000:167-189. DOI: 10.1007/978-1-62703-405-0_13

[40] Meng Z, Veenstra TD. Targeted mass spectrometry approaches for protein biomarker verification. J. Proteomics 2011;74:2650-2659. DOI: 10.1016/j.jprot.2011.04.011

[41] Ma H, Chen G, Guo M. Mass spectrometry based translational proteomics for biomarker discovery and application in colorectal cancer. Proteomics Clin. Appl 2016;10:503-515. DOI: 10.1002/prca.201500082

[42] Gerszten RE, Carr SA, Sabatine M. Integration of proteomic-based tools for improved biomarkers of myocardial injury. Clin. Chem 2010;56:194-201. DOI: 10.1373/clinchem.2009.127878

[43] Zhou J, Yin Y. Strategies for large-scale targeted metabolomics quantification by liquid chromatography-mass spectrometry. Analyst 2016;141:6362-6373. DOI: 10.1039/c6an01753c

[44] Kuzyk MA, Parker CE, Domański D, Borchers CH. Development of MRM-based assays for the absolute quantitation

of plasma proteins. Methods Mol. Biol 2013;1023:53-82. DOI: 10.1007/978-1-4614-7209-4_4

[45] Bill RM, Ulke-Lemée A, Cartwright CA, Far R, Kim J, MacDonald JA. Analyzing recombinant protein production in Pichia pastoris with targeted proteomics. Methods Mol. Biol 2019;1923:187-209. DOI: 10.1007/978-1-4939-9024-5_8

[46] Mohammed Y, Domański D, Jackson AM, Smith DS, Deelder AM, Palmblad M, Borchers CH. PeptidePicker: a scientific workflow with web interface for selecting appropriate peptides for targeted proteomics experiments. J. Proteomics 2014;106:151-161.

[47] Abbatiello SE, Schilling B, Mani DR, Zimmerman LJ, et al. Large-scale interlaboratory study to develop, analytically validate and apply highly multiplexed, quantitative peptide assays to measure cancer-relevant proteins in plasma. Mol. Cell. Proteomics. 2015; 14:2357-2374. DOI: 10.1074/mcp.M114.047050

[48] Aguilar-Mahecha A, Kuzyk MA, Domanski D, Borchers CH, Basik M. The effect of pre-analytical variability on the measurement of MRM-MS-based mid- to high-abundance plasma protein biomarkers and a panel of cytokines. PLoS One 2012;7:e38290. DOI: 10.1371/journal.pone.0038290

[49] Percy AJ, Yang J, Hardie DB, Chambers AG, Tamura-Wells J, Borchers CH. Precise quantitation of 136 urinary proteins by LC/MRM-MS using stable isotope labeled peptides as internal standards for biomarker discovery and/or verification studies. Methods 2015;81:24-33. DOI: 10.1016/j.ymeth.2015.04.001

[50] Catalona WJ, Smith DS, Ratliff TL, Dodds KM, Coplen DE, Yuan JJ, Petros JA, Andriole GL. Measurement of prostate-specific antigen in serum as a screening test for prostate cancer. N. Engl. J. Med 1991;324:1156-1161. DOI: 10.1056/NEJM199104253241702

[51] Jimenez RV, Szalai AJ. Therapeutic lowering of C-reactive protein. Front. Immunol 2021;11:619564. DOI: 10.3389/fimmu.2020.619564

[52] Xu X, Keefer LK, Zeigler R, Veenstra TD. A liquid chromatography-mass spectrometry method for the quantitative analysis of urinary endogenous estrogen metabolites. Nat. Protoc 2007;2:1350-1355. DOI: 10.1038/nprot.2007.176

[53] Li L, Wang L, Shangguan D, Wei Y, Han J, Xiong S, Zhao Z. Ultra-high-performance liquid chromatography electrospray ionization tandem mass spectrometry for accurate analysis of glycerophospholipids and sphingolipids in drug resistance tumor cells. J. Chromatogr. A 2015;1381:140-148. DOI: 10.1016/j.chroma.2015.01.013

[54] Gaither C, Popp R, Mohammed Y, Borchers CH. Determination of the concentration range for 267 proteins from 21 lots of commercial human plasma using highly multiplexed multiple reaction monitoring mass spectrometry. Analyst 2020;145:3634-3644. DOI: 10.1039/c9an01893j

[55] Anderson NL. The clinical plasma proteome: a survey of clinical assays for proteins in plasma and serum. Clin. Chem 2010;56:177-185. DOI: 10.1373/clinchem.2009.126706

[56] Domanski D, Percy AJ, Yang J, Chambers AG, Hill JS, Freue GV, Borchers CH. MRM-based multiplexed quantitation of 67 putative cardiovascular disease biomarkers in human plasma. Proteomics 2012;12:1222-1243. DOI: 10.1002/pmic.201100568

[57] Link AJ. Multidimensional peptide separations in proteomics. Trends Biotechnol 2002;20:S8-13. DOI: 10.1016/s1471-1931(02)00202-1

[58] Anderson NL, Anderson NG. The human plasma proteome: history, character, and diagnostic prospects. Mol. Cell. Proteomics 2002;1:845-867. DOI: 10.1074/mcp.r200007-mcp200

[59] Matsui NM, Borsig L, Rosen SD, Yaghmai M, Varki A, Embury SH. P-selectin mediates the adhesion of sickle erythrocytes to the endothelium. Blood 2001;98:1955-1962. DOI: 10.1182/blood.v98.6.1955

[60] Ataga KI, Kutlar A, Kanter J, Liles D, Cancado R, Friedrisch J, Guthrie TH, Knight-Madden J, Alvarez OA, Gordeuk VR, Gualandro S, Colella MP, Smith WR, Rollins SA, Stocker JW, Rother RP. Crizanlizumab for the prevention of pain crises in sickle cell disease. N. Engl. J. Med 2017;376:429-439. DOI: 10.1056/NEJMoa1611770

[61] Novartis wins early FDA approval of P-selectin binder Adakveo in SCD [Internet] 2021. Available from: https://www.bioworld.com/articles/431143-novartis-wins-early-fda-approval-of-p-selectin-binder-adakveo-in-scd?v=preview [Accessed: 2021-05-30]

[62] De Buck M, Gouwy M, Wang JM, Van Snick J, Opdenakker G, Struyf S, Van Damme J. Structure and expression of different serum amyloid A (SAA) variants and their concentration-dependent functions during host insults. Curr. Med. Chem. 2016;23:1725-1755. DOI: 10.2174/0929867323666160418114600.

[63] MTM Proteomics PeptiQuant™ Plus Kits [Internet]. 2021. Available from: https://shop.isotope.com/category.aspx?id=10032893 [Accessed: 2021-06-02]

[64] Lv PC, Yang YS, Wang ZC. Recent progress in the development of small molecule c-Met inhibitors. Curr. Top. Med. Chem 2019;19:1276-1288. DOI: 10.2174/1568026619666190712205353

[65] Van Cutsem E, Karaszewska B, Kang YK, Chung HC, Shankaran V, Siena S, Go NF, Yang H, Schupp M, Cunningham D. A multicenter phase II study of AMG 337 in patients with Met-amplified gastric/gastroesophageal junction/esophageal adenocarcinoma ad other Met-amplified solid tumors. Clin. Cancer Res 2019;25:2414-2423. DOI:10.1158/1078-0432.CCR-18-1337

[66] Catenacci DV, Henderson L, Xiao SY, Patel P, Yauch RL, Hegde P, Zha J. Durable complete response to metastatic gastric cancer with anti-Met therapy followed by resistance at recurrence. Cancer Discov 2011;1:573-579. DOI: 10.1158/2159-8290.CD-11-0175

[67] Doshi S, Gisleskog PO, Zhang Y, Zhu M, Oliner KS, Loh E, Perez Ruixo JJ. Rilotumumab exposure-response relationship in patients with advanced or metastatic gastric cancer. Clin. Cancer Res 2015;21:2453-2461. DOI: 10.1158/1078-0432.CCR-14-1661

[68] Neagu AN. Proteome imaging: from classic to modern mass spectrometry-based molecular histology. Adv. Exp. Med. Biol 2019;1140:55-98. DOI: 10.1007/978-3-030-15950-4_4

[69] van Bogaert LJ. P16INK4a immunocytochemistry/immunohistochemistry: need for scoring uniformization to be clinically useful in gynecological pathology. Ann. Diag. Pathol 2012;16:422-426. DOI: 10.1016/j.anndiagpath.2012.03.006

[70] Hood BL, Darfler MM, Guiel TG, Furusato B, Lucas DA, Ringeisen BR, Sesterhann IA, Conrads TP, Veenstra TD,

Krizman DB. Proteomic analysis of formalin-fixed prostate cancer tissue. Mol. Cell. Proteomics 2005;5:1741-1753. DOI: 10.1074/mcp.M500102-MCP200

[71] Catenacci DV, Liao WL, Thyparambil S, Henderson L, Xu P, Zhao L, Rambo B, Hart J, Xiao SY, Bengali K, Uzzell J, Darfler M, Krizman DB, Cecchi F, Bottaro DP, Karrison T, Veenstra TD, Hembrough T, Burrows J. Absolute quantitation of Met using mass spectrometry for clinical application: assay precision, stability, and correlation with MET gene amplification in FFPE tumor tissue. PLoS One 2014;9:e100586. DOI: 10.1371/journal. Pone.0100586

[72] Catenacci DVT, Liao WL, Zhao L, Whitcomb E, Henderson L, O'Day E, Xu P, Thyparambil S, Krizman D, Bengali K, Uzzell J, Darfler M, Cecchi F, Blackler A, Bang YJ, Hart J, Xiao SY, Lee SM, Burrows J, Hembrough T. Mass-spectrometry-based quantitation of Her2 in gastroesophageal tumor tissue: comparison to IHC and FISH. Gastric Cancer 2016;19:1066-1079. DOI: 10.1007/ s10120-015-0566-0

[73] Do M, Kim H, Yeo I, Lee J, Park IA, Ryu HS, Kim Y. Clinical application of multiple reaction monitoring-mass spectrometry to human epidermal growth factor receptor 2 measurements as a potential diagnostic tool for breast cancer therapy. Clin. Chem 2020;66:1339-1348. DOI: 10.1093/ clinchem/hvaa178

[74] Hart V, Gautrey H, Kirby J, Tyson-Capper A. HER2 splice variants in breast cancer: investigating their impact on diagnosis and treatment outcomes. Oncotarget. 2020;11:4338-4357. DOI: 10.18632/oncotarget.27789

[75] Lehmann WD. A timeline of stable isotopes and mass spectrometry in the life sciences. Mass Spectrom. Rev. 2017;36:58-85. DOI: 10/1002/mas.21497

[76] Paterson RW, Gabelle A, Lucey BP, Barthélemy NR, Leckey CA, Hirtz C, Lehmann S, Sato C, Patterson BW, West T, Yarasheski K, Rohrer JD, Wildburger NC, Scott JM, Karch CM, Wray S, Miller TM, Elbert DL, Zetterberg H, Fox NC, Bateman RJ. SILK studies – capturing the turnover of proteins linked to neurodegenerative diseases. Nat. Rev. Neurol 2019;15:419-427. DOI: 10.1038/s41582-019-0222-0

[77] Bateman RJ, Munsell LY, Chen X, Holtzman DM, Yarasheski KE. Stable isotope labeling tandem mass spectrometry (SILT) to quantify protein production and clearance rates. J. Am. Soc. Mass Spectrom 2007;18:997-1006. DOI: 10.1016/j.jasms.2007.02.009

[78] Bonten M, Johnson JR, van den Biggelaar AHJ, Georgalis L, Geurtsen J, de Palacios PI, Gravenstein S, Verstraeten T, Hermans P, Poolman JT. Epidemiology of Escherichia coli bacteremia: a systematic literature review. Clin. Infect. Dis 2021;72:1211-1219. DOI: 10.1093/cid/ciaa210

[79] Heming N, Azabou E, Cazaumayou X, Moine P, Annane D. Sepsis in the critically ill patient: current and emerging management strategies. Expert Rev. Anti. Infect. Ther 2021;19:635-647. DOI: 10.1080/14787210.2021.1846522

[80] Dela Cruz WP, Richardson JY, Broestler JM, Thornton JA, Danaher PJ. Rapid determination of macrolide and lincosamide resistance in group B streptococcus isolated from vaginal-rectal swabs. Infect. Dis. Obstet. Gynecol 2007;2007:46581-46587. DOI: 10.1155/2007/46581

[81] Vallés J, Rello J, Ochagavía A, Garnacho J, Alcalá MA.

Community-acquired bloodstream infection in critically ill adult patients: impact of shock and inappropriate antibiotic therapy on survival. Chest 2003;123:1615-1624. DOI: 10.1378/chest.123.5.1615

[82] Wang X, Ma L, Sun S, Liu T, Zhou H, Liu X, Guan M. Rapid, highly sensitive and quantitative detection of interleukin 6 based on SERS magnetic immunoassay. Anal. Methods 2021;13:1823-1831. DOI: 10.1093/d0ay02304c

[83] Cantey JB, Lee JH. Biomarkers for the diagnosis of neonatal sepsis. Clin. Perinatol 2021;48:215-227. DOI: 10.1016/j.clp.2021.03.012

[84] Anhalt JP, Fenselau C. Identification of bacteria using mass spectrometry. Anal. Chem 1975;47:219-225. DOI: 10.1021/ac60352a007

[85] Cain TC, Lubman DM, Weber WJ Jr. Differentiation of bacteria using protein profiles from matrix-assisted laser desorption/ionization time-of-flight mass spectrometry. Rapid Commun. Mass Spectrom 1994;8:1026-1030. DOI: 10.1002/rcm.1290081224

[86] Claydon MA, Davey SN, Edwards-Jones V, Gordon DB. The rapid identification of intact microorganisms using mass spectrometry. Nat. Biotechnol 1996;14:1584-1586. DOI: 10.1038/nbt1196-1584

[87] Nomura F, Tsuchida S, Murata S, Satoh M, Matsushita K. Mass spectrometry-based microbiological testing for blood stream infection. Clin. Proteomics 2020;17:14-24. DOI: 10.1186/s12014-020-09278-7

[88] Broyer P, Perrot N, Rostaing H, Blaze J, Piston F, Gervasi G, Charles MH, Dachaud F, Dachaud J, Moulin F, Cordier S, Dauwalder O, Meugnier H, Vandenesch F. An automated sample preparation instrument to accelerate positive blood cultures microbial identification by MALDI-TOF mass spectrometry (Vitek MS). Front. Microbiol 2018;9:911-924. DOI: 10.3389/fmicb.2018.00911

[89] Ryzhov V, Fenselau C. Characterization of the protein subset desorbed by MALDI from whole bacterial cells. Anal. Chem 2001;73:746-750. DOI: 10.1021/ac0008791

Chapter 5

Interrelation between Recurrent Pregnancy Loss and Antiphospholipid: A Clinico-Diagnostic Perspective

Nilam Bhasker

Abstract

Recurrent pregnancy loss, one of the crucial reproductive health concerns affecting 6% of couples. Clinically recognized pregnancy loss is familiar, occurring in approximately 15–25% cases of pregnancy. The most common cause of recurrent pregnancy loss is cytogenetic anomaly, antiphospholipid antibody, metabolic and hormonal disorders. However, approximately 50% cases of recurrent pregnancy loss remain unexplored. Recurrent pregnancy loss is correlated with specificity of antiphospholipid like anti-β2-glycoprotein-I antibodies, lupus anticoagulant, anti-cardiolipin antibodies, and anti-phosphatidylserine. aPL inhibits the release of human chorionic gonadotropin (HCG) hormone from placenta, trophoblast growth, migration, and cell adhesion while induce the inflammatory response in earlier pregnancy. Some clinical studies reported that occurrence of antiphospholipid during recurrent pregnancy loss is uncommon. In this time line article, we are focusing on the role of antiphospholipid in the recurrent pregnancy loss and clinico-diagnostic against recurrent pregnancy loss.

Keywords: Recurrent pregnancy loss, Cytogenetic anomaly, Antiphospholipid and Diagnosis

1. Introduction

Premature loss of pregnancy referred to spontaneous abortion or miscarriage is described as loss of clinical pregnancy after the 18 weeks of fertilization (gestational age-20 weeks). Guidelines from ESHRE (the European Society of Human Reproduction and Embryology) and ASRM (the American Society for Reproductive Medicine) interpret recurrent pregnancy loss or miscarriage was traditionally described as consecutive loss of 3 or more pregnancy [1, 2] prior to 24 weeks of gestational age and molar (characterized by superfluous placental growth owing to an atypical fertilization of egg) and ectopic (implantation of embryo outward the uterine cavity) pregnancies are not incorporated in definition set by ESHRE and ASRM. It is a frequent obstetric complication prevailing in approximately 15–25% of clinically recognized pregnancy [2] which is confirmed by histopathology or ultrasonography. Moreover, epidemiological studies have shown that the frequency of succeeding loss of pregnancy is 24% post 2 pregnancy losses while it is 30% and 40% after 3 and 4 subsequent pregnancy losses respectively [3, 4]. The prevalence of obstetric complications increases with maternal age (**Figure 1**) [5]. The patho-physiology of recurrent miscarriage differs according to age of gestation and maternal, however several mechanisms may merged at some points that induces the loss of pregnancy. The most familiar mechanisms are chromosomal abnormalities in the conceptus which inhibit the further development and disintegrate the interface between the fetus and maternal resulting in cramping, bleeding, and miscarriage. Several etiological factors have been associated with recurrent miscarriage but they influence only few cases of pregnancies which is still not clear. Approximately in case of more than 50% of women, the risk factor is not determined for pregnancy loss [6, 7]. Incidence of premature pregnancy loss is depending upon the method used by clinician to recognize the pregnancy. Approximately 50% of all conceptions are found to lost at pre-clinical stages due to failure of implantation or biochemical loss [1, 2, 8, 9] while clinically recognized pregnancies loss are reported in 9–20% of cases [9, 10], especially during 1st trimester (gestational age: 5–12 weeks) [11]. Week by week, the rate of miscarriage is varied during early pregnancy. One of the studies showed a sharp decline in the incidence rate by ~1% post twelve weeks of gestation [11] suggesting that in most of the cases the loss of pregnancy takes place shortly after implantation. Diagnosis of miscarriage is depend upon self-reporting (delayed menstrual cycle and pregnancy test at home) and clinical testing including histopathology, trans-vaginal ultrasonography and declining level of serum human chorionic gonadotrophin (hCG). Distinction must be made between embryonic, fetal losses and biochemicals because pregnancy loss is varied according to gestational age. Nonetheless, the prognostic utility of staging 1st trimester pregnancy losses on the basis of gestational age is not clear, hence it is not reported in the studies. Basically, discrimination between the 1st and 2nd trimester pregnancy loss is usually done in clinical practice. In most

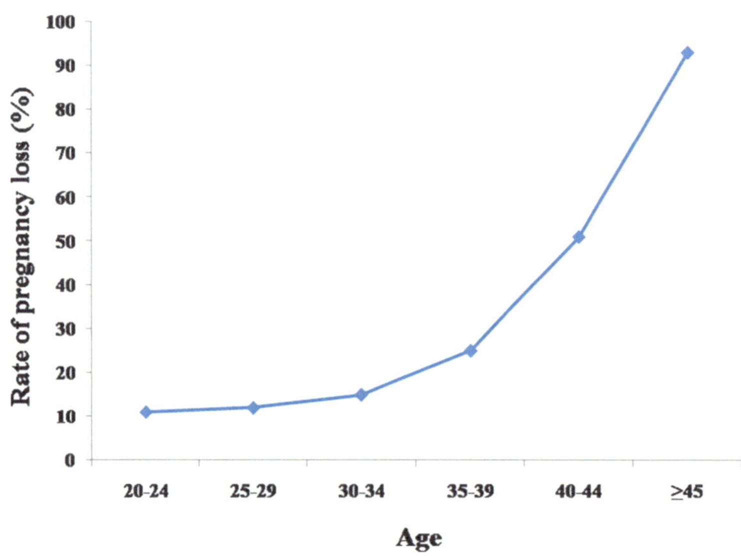

Figure 1.
Prevalence of recurrent pregnancy loss increases with age.

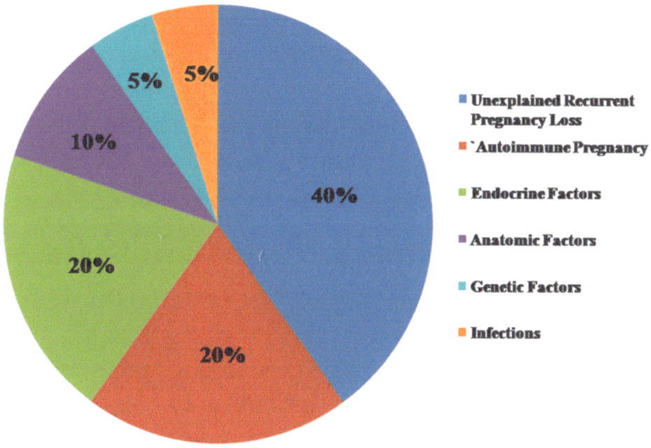

Figure 2.
Etiological factors of recurrent pregnancy loss.

of the cases, risk factor (etiology) is not clear but several studies recognized the risk factors including uterine anatomic and chromosomal anomalies, endocrine anomalies, infections of endometrial, alloimmune causes, inherited thrombophilias, stress factors, exposure to environmental factors, genetic factors and antiphospholipid syndrome (**Figure 2**) [4, 12–14]. Several studies focused on the relationship between the presence of antiphospholipid and recurrent pregnancy loss and they reported the positive relationship [15–17]. A study conducted by Rai et al., [18] showed that fetal loss (≥10 weeks of gestation) is robustly linked with antiphospholipid in contrast to earlier pregnancy losses. In this review, we will discuss the clinic-pathological approach and analysis of the pathogenic association between recurrent pregnancy loss and antiphospholipid.

2. Pathogenic role: antiphospholipid

Antiphospholipid triggers the obstetric complications (or thrombosis) by influencing the functional aspect of various cells including monocytes, platelets and endothelial cells on the vascular side, decidual and trophoblast cells and neutrophils during the 1st trimester of pregnancy. Antiphospholipid interacts with these targets through multiple receptors including Toll-like Receptor (TLR)-9, TLR2, TLR4, heparan-sulphate, Annexin A2, glycoprotein (GP)-Iba and apolipoprotein E (APOE) receptor-20. On interaction with receptor, antiphospholipid activates the multiple intracellular mediators like PI3K (phosphatidylinositol 3-kinase)-AKT pathway, p38-MAPK (mitogen-activated protein kinase), NFk-B (nuclear factor k-B) resulting in the recruitment of mTOR and NLRP3 as well as Caspase-1 by the activation of NOX-2 (endosomal NADPHoxidase) [19]. Eventually, these signaling events induce the overexpression of various pro-inflammatory molecules (TNFα, IL-6, IL-8 and tissue factor). In spite of this, antiphospholipid interacts with various soluble molecules of coagulation and also activates the complement system. One of the *in vivo* studies showed that antiphospholipid induces the pro-thrombotic pathogenic potential in animals that are earlier challenged with lipopolysaccharide (LPS) or mechanical or photochemical trauma [20]. Antiphospholipid IgG is efficient to trigger the fetal loss in mice with pregnancy without 2nd hit which is necessary for occurrence of blood clotting for thrombophilic condition [20]. This might be owing to high expression of β2-glycoprotein-I (β2-GPI) at the physiological level. Decidual and trophoblast cells express anionic phospholipid on the cell membrane like phosphatidylserine which acts as an anchor for β2-GPI [20, 21]. An *in vitro* study conducted by Poulton et al., [22] showed that immunoglobulins from obstetric and thrombotic APS patients elicit different biological response because immunoglobulins extracted from obstetric patients impede the invasion of trophoblast in contrast to immunoglobulins extracted from non-obstetric patients. Antiphospholipid inhibits the progress of pregnancy by damaging the both ends of tissue: (1) Embryo side: Antiphospholipid disrupts this end by various ways; (a) impaired the trophoblast invasion through inducing the inhibitors of MMP (matrix metalloproteinase) resulting in down-regulation of MMP; (b) affect the differentiation of trophoblast by reducing the secretion of β-hCG; (c) promotes the apoptosis of trophoblast; (d) activation of conventional complement system that leads to synthesis of TNF-α culminating with recruitment of inflammatory molecules [20, 23, 24]. (2) Maternal side: impaired the endometrial angiogenesis by disrupting the ratio of angiogenic factors. Hence, there are 3 key mechanisms associated with pathogenesis of obstetric APS i.e. inflammation, complement activation and placental thrombosis. Antiphospholipid triggers the focal thrombosis at the terminal end of uteroplacental vascularization which culminates with placental infarction and impaired the blood exchange between maternal and fetus. In spite of this, antiphospholipid induces the placental thrombosis by disturbing the Annexin on epithelial and trophoblast cells that acts as A5 anticoagulant shield. One of the studies performed on antiphospholipid-positive placenta showed the decreased dispersion of Annexin A5 on the surface of intervillous [25]. Placental infarction occurs chiefly owing to an intra-luminal thrombus that occludes the spiral artery and it is a familiar histo-pathological manifestation in obstetric APS. Although few studies demonstrated that spiral artery thrombosis or placental infarction is not observed in the placenta isolated from 1st trimester abortions [26, 27] and suggested that placental infarction or spiral artery thrombosis might be implicated only for complications that occur during the late stage of gestational period [21] (**Figure 3**).

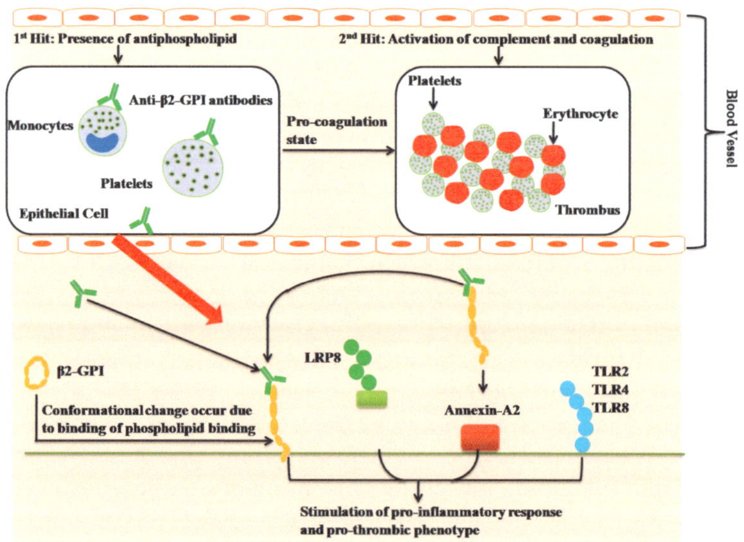

Figure 3.
Mechanisms of recurrent pregnancy loss.

3. Clinical manifestation: antiphospholipid syndrome (APS)

Antiphospholipid syndrome (APS), an autoimmune disorder in which auto-antibodies are developed against self molecules in the individuals such as in case of APS, auto-antibodies (antiphospholipid) binds with phospholipid binding proteins. APS is a causative agent of recurrent pregnancy loss [28] and a main clinical manifestation is thrombosis or recurrent pregnancy loss. In spite of this, renal impairment, thrombocytopenia, livedo reticularis and vulvular heart diseases. Approximately 10–15% of women with recurrent pregnancy loss are detected with APS [29, 30]. One of the studies showed that it occurs in approximately 38.6% of European women with APS [31]. van Dijk et al., [7] demonstrated that similar prevalence is found in both APS and women with 2 or women with 3 or more losses. The precise mechanism of recurrent pregnancy loss associated with APS is not clear. Several lines of evidence suggested that detrimental inflammatory response by autoantbodies on endothelial cells and placenta in contrast to thrombosis which is initially assumed for recurrent pregnancy loss [32–34]. Some of the studies based on humans demonstrated that excessive complement activation is responsible for early pregnancy loss [35]. Preclinical studies showed that antiphospholipid activates the complement resulting in recurrent pregnancy loss [36]. Another study supports this hypothesis as treatment with anti-C5 antibody prevents the fetal loss in pregnant mice injected with antiphospholipid antibody [37]. Thrombosis and recurrent pregnancy loss are

1. **Thrombosis:** It is the formation of blood clots inside the blood vessels which may be micro-vascular, venous and artery; may be all of these present in the same patients.

2. **Recurrent Pregnancy Loss:** Recurrent pregnancy loss is described as loss of 3 or more pregnancy. It excludes the ectopic and molar pregnancy and is the main feature in case of obstetrics.

4. Clinical approach

The preliminary role of this approach is diagnosis and treatment of disease in individuals.

4.1 Diagnosis

Assessment and management of recurrent miscarriage must include the recognition of modifiable risk factor and avoid the invasive testing for perceived risk factor [38]. Multinational prospective studies are needed to improve the potency of diagnostic test especially in case of live birth rates in women with recurrent miscarriage.

4.1.1 Diagnosis: pregnancy loss

Reduction in the level of hCG hormone in the 1st trimester is indication of pregnancy loss, but further additional examination is required like histo-pathological examination (identification of placental villi). ELISA (enzyme linked immunosorbent assay) used to determine antiphospholipid and it is a popular method for detection of antiphospholipid by using purified antigens. ELISA is useful in evaluating the recurrent pregnancy loss by determining the antinuclear positive antibodies, especially to prevent the recurrent pregnancy loss during the 1st trimester. Basically, a solid phase Immunoassay is carried out on cardiolipin-coated plate (different coating for o detect the different antibody) usually in the presence of blocking agent (fetal calf or bovine serum) that consists of β2-glycoprotein-I which acts as antigenic epitope to anti-cardiolipin antibody and eventually leads to detection of antibody [39]. In spite of this, other diagnostic method include screening of POC (products of conception) and ultrasonography [1]. Trans-vaginal ultrasonography is considered as a gold standard for the testing of pregnancy loss (**Table 1**) while the exact diagnostic criteria are still under debate [40]. Obstetricians and Gynecologists of different countries have different setting associated with mean sac diameter (MSD) like in the USA; they consider the bare gestation sac with ≥16 mm MSD or CRL (crown rump length) with ≥5 mm and absence of heartbeat while MSD with ≥20 mm or CRL with 6 mm in the UK for diagnosis of pregnancy loss [41, 42]. False positive diagnosis may occur during the test of pregnancy due to this variability; hence, a strict criterion has been developed by the international committee of Obstetricians and Gynecologists i.e. MSD with ≥25 mm or CRL with ≥7 mm and absence of heartbeat [43]. This criterion has been confirmed by large multicentre prospective observational studies and also set criteria for follow up scan to verify the pregnancy loss like absence of heartbeat after 7 days of 1st scan of embryo with ≤7 mm CRL and failing to attain the doubled size in ≥14 days for vacant gestational sac with <12 mm MSD. In spite of this, estimation of concentration of progesterone hormone provides more information regarding the pregnancy loss [43]. Less than 10 μg/l concentration of progesterone hormone represents the robust indication of pregnancy loss [44]. In 2012, ASRM recommended the guidelines for absolute assessment of pregnancy loss in women having 2 clinically recognized pregnancy losses including evaluation of uterine anatomy, karyotyping of both parents, assessment of prolactin and thyroid hormone abnormalities and APS. ESHRE provide more extensive guidelines with a graded system for diagnosis of pregnancy loss that includes evaluation of the uterine cavity, APS and screening of thyroid. They also recommend the conditional guidelines for parental karyotyping and prolactin anomalies. Karyotyping diagnosis is used when couples were having enhanced risk of aneuploidy or couples having family history for abnormal translocation of chromosome or in case of congenital anomalies [1]. Contrarily, diagnostic test

Investigation	Rational
Antiphospholipid antibodies (lupus anticoagulant & IgM and IgG anti-cardiolipin)	Approximately 15% of women have APS (antiphospholipid syndrome). Treatment with low dose aspirin and low molecular weight heparin significantly improves the outcomes of live-birth rate.
Parental karyotyping where testing POC (products of conception) which reports an unbalanced chromosomal anomaly	Whenever parental karyotype is unusual, there is a stronger prognosis for the following pregnancy. If unbalanced chromosomal abnormality perform parental karyotypes. Approximately 4% of couples have an unusual karyotype. Recommend for genetic counseling.
Trans-vaginal pelvic ultrasound scan	To determine ovarian morphology and uterine anatomy. Suspected uterine abnormalities may need further examination by using laparoscopy, hysteroscopy or 3D-pelvic ultrasound.

Table 1.
Diagnosis of recurrent pregnancy loss.

for prolactin anomalies are used when women possess manifestation of hyperprolactinaemia incuding women with amenorrhoea or oligomenorrhoea [1]. Both ESHRE and ASRM did not recommend the screening of thrombophilia and recommend thrombophilia screening only in case of individuals having family history of thrombophilia [1, 2]. Additionally, ESHRE did not also recommend the routine testing of chronic endometritis, POC, ovarian reserve and luteal insufficiency [1].

1. **Screening of APS:** Clinical guidelines recommended the identification of antiphospholipid antibodies including IgM & IgG anti-β2-glycoprotein-I, lupus anticoagulant and IgM & IgG anti-cardiolipin for the detection of APS in women with recurrent miscarriage (**Table 1**). Screening of anti-β2-glycoprotein-I is not performed in some countries including UK because weak association has been reported between anti-β2-glycoprotein-I and recurrent pregnancy loss as well as it provides less predictive outcomes [45]. APS is manifested by the existence of one laboratory and one clinical criterion:

    I. Laboratory (replicated at the minimum of 2 times, >12 weeks apart);

    - +ve plasma level (high to medium) of anti-cardiolipin

    - +ve plasma level of lupus anti-coagulant

    II. Clinical;

    - More than one established episodes of vascular thrombosis (Arterial, Venous and small vessel)

    - Obstetric complications including 3 or more subsequent loss of pregnancy at <10 weeks of gestational age, more than one fetal demise at >10 weeks of gestational age or premature birth at <34 weeks owing to placental insufficiency or dreadful pre-eclampsia

2. **Parental karyotyping and cytogenetic analysis of POC:** Reason behind the conducting the regular parental karyotyping owing to existence of a Robertsian translocation (chromosomal inversion, supernumerary chromosome or sex chromosome aneuploidy) or balanced reciprocal translocation which is accounts for 2–5% of couples with recurrent miscarriage (**Table 1**) [46–48]. According to ASRM guidelines, parental karyotyping provides valuable information about prognosis of succeeding pregnancies [10]. ESHRE clinical guidelines underline the successful collective live births even they have chromosomal anomalies, hence probability of recurrent miscarriage are greater but the probability of live birth is very low when the fetus have balanced translocation [1, 49]. ASRM notified the barrier associated with G-banding karyotyping examination i. e. adulteration of maternal tissue like in case of cell culture based analysis of G-banding karyotyping where probability of maternal contamination is high including selective expansion of mosaic cells or adulteration of 46XX embryos [50, 51]. Issues with adulteration of maternal tissue can be overcome CMA (24-chromosomal microarray) diagnostic testing of maternal blood. Although assays having greater sensitivity may recognize variants which have no impact on

pregnancy loss or recognize the mosaicism that are restricted to placental cells [10, 52]. Hence, CMA is considered as gold standard test for the genetic evaluation of POC [51, 53]. One of the studies conducted on 22,451 miscarriage samples showed that approximately 59% of miscarriage samples possess chromosomal anomalies [51]. Additionally, CMA test has some limitations including detection of lesser number of mosaicism (<10–15%) and unable to identify the balanced structural rearrangement of chromosome that may promote the early pregnancy loss [54, 55]. A prospective study conducted on 100 women by using both CMA test and ASRM guidelines observed the probable clarification of pregnancy loss in approximately of 90% of women [56, 57]. Although, genetic testing of POC is not promising in case of outcomes in women with recurrent miscarriage while it only provide the reason for the pregnancy loss. Future numerous studies are needed to verify that regular genetic testing associated with POC must be an intrinsic part of psychological and clinical management of women with recurrent miscarriage.

3. **Diagnostic Test for hormone anomalies:** Assessment of prolactin in serum is not recommended for women with recurrent miscarriage except they possess the clinical symptoms for hyper-pro-lactinaemia [1, 10]. Screening and testing of thyroid function for the existence of thyroid peroxidase antibodies and TSH (thyroid stimulating hormone) are recommended for women who have recurrent miscarriage [1].

4. **Evaluation of the uterine cavity:** Estimation of the uterine cavity for acquired and congenital abnormalities is basically performed by using 3D-ultrasonography (**Table 1**) [1, 10]. MRI, saline-infused sonography or hysteroscopy can also be utilized particularly in case of unavailability of 3D-ultrasonography.

4.1.2 Diagnostic test

There are two broad tests to diagnose the APS (antiphospholipid antibodies):

1. **ELISA:** This is a biochemical assay to detect the ligand (more commonly protein) and it is employed to determine the antiphospholipid antibodies. Two different ELISAs (anti-2GPI and the anti-cardiolipin) are employed to diagnose the APS [57]. In this method, ELISA plate is coated with either the anionic phospholipid cardiolipin or 2GPI followed by adding the pre-specified diluted (1:50) serum of the patient. Further, add the secondary labeled antibody which helps to quantify the bound IgM or IgG isotypes [58]. The anti-cardiolipin ELISA recognizes the antibody that specifically binds to cardiolipin as well as other anionic phospholipids. Anti-2GPI ELISA recognizes the antibody that interacts with 2GPI which is coated on the surface on anionic phospholipid [59]. The anti-cardiolipin ELISA is less specific in contrast to anti-2GPI ELISA to diagnose the APS as it also detects non-specific antibodies present in the serum due to various infections [60]. To overcome this problem, retesting is recommended for antiphospholipid antibodies after 12 weeks of initial tests [57] and acts as a safety precaution against false positive readings. The advantages of ELISA to determine the antiphospholipid antibody are as follows: (1) Several samples can be analyzed at the same time, (2) It is independent of anticoagulant treatment and coagulation confounding factors.

4.1.3 Principle

In the standard for aCL or β2GPI ELISA, polystyrene plate is coated with cardiolipin/β2GPI in ethanol, and then dried the plate through evaporation of ethanol by placing the plate to air. Plate is blocked with a blocking agent (10% fetal calf serum, or 1% BSA) for 1–2 hours after washing the plate with PBS. Plate is again washed with PBS and then adds the patient's sample at the dilution of 1/50–1/100. Sufficient number of standards must be taken to generate the standard calibration plot / curve. Microtitre plate is incubated for 2 to 3 hours; after incubation, the plate is washed again with PBS. Labeled (alkaline phosphatase) anti-human IgM or IgG is added to the plate at the appropriate dilution followed by incubation 1to 2 hours. The substrate (p-nitrophenyl phosphate) is added. 3 N NaOH is employed to prevent the reactions especially when standard calibration curve is reached to pre-determined OD (optical density). The estimation of aCL/β2GPI levels is determined from the standard calibration curve [61].

1. **Lupus anticoagulant assay (LAC):** There are several assays to perform the LAC assay in laboratories to detect the LAC including dRVVT (dilute Russell Viper Venom Test), dPT (dilute Prothrombin Time), STACLOT-LA (Hexagonal Phase Phospholipid Neutralization), KCT (Kaolin Clotting time), SCT (Silica Clotting Time), Tissue Thromboplastin Inhibition, and Platelet Neutralization Procedure. There are two other less commonly employed assays i. e. the Textarin:Ecarin Ratio and Taipan Venom Time [62]. Single assay is not enough to identify the all LAC antibody owing to the heterogeneity nature of LAC antibody. Hence, it is recommended that laboratories should be performing at least two assays to test the LAC antibody. LAC assay is based on coagulation and measures the *in vitro* clotting time of the plasma of a patient with respect to control samples which is called a screening test. Two further diagnosing test performed after prolongation time for clotting is noted: (1) To estimate the prolonged time for clotting is reversed after mixing (1:1 ratio) the plasma of patient with pooled plasma from normal healthy controls, (2) To estimate the prolongation of clotting time (observed in screening test) is reversed after the inclusion of surplus amount of anionic phospho-lipid. It is considered a confirmatory test. These tests are collectively known as LAC assays. Plasma sample is considered as positive (+) for LAC when prolonged coagulation time is not reversed after the mixing anionic phospho-lipid while it is reversed after addition of excess phospho-lipid [63]. Venom extracted from Russel viper is employed in the dRVVT assay in combination with dilute phosphor-lipids to trigger the activation of factor X. KCT, aPTT, and SCT assays stimulate the contact activation and they are compatible with automated analyzers [64].

4.2 Treatment: women possess +ve antiphospholipid

Identification of antiphospholipid has been progressively carried out in asymptomatic women, particularly for obstetrical reasons. In parallel, low titer detection of antiphospholipid is very common-the concurrence benchmark for APS do not comprise sero-positive anti-β2-GPI and anti-cardiolipin beneath the 99%-in patients with obstetric characteristics. Hence, it is not uncommon to grasp the authority to recommend treatment throughout the gestation period in healthy women with antiphospholipid.

1. **Low dose of aspirin (LDASA):** LDASA is a primary prophylaxis drug for the obstetric and also advisable in case of women without antiphospholipid to prevent the pre-eclampsia [58]. Several physicians are utilizing this drug to control the pregnant carriers of antiphospholipid especially in patients who have earlier experienced the 1 or 2 fetal loss or in case of coexistence of maternal risk factors like obesity, arterial hypertension, and age [59]. A macro-scale retrospective observational study demonstrated that outcomes of pregnancy may be beneficial without LDASA particularly in case of patients carrying low-risk antiphospholipid profile [60]. Another large cohort study performed on 73 pregnant antiphospholipid carriers (largely isolated lupus anticoagulant), LDASA treatment yielded a favorable pregnancy outcome which was equivalent to the control population [61]. A study carried out at multicenter by collecting 200 women carrying +ve antiphospholipid during pregnancy showed that approximately 18% of antiphospholipid carriers have adverse pregnancy outcome, similarly in case of thrombotic APS (24%) and obstetric APS (18%) [62]. Triple antiphospholipid sero-positivity is the only risk factor for adverse outcomes associated with pregnancy [62]. Further studies suggested that additional treatments (low molecular weight heparin or immunemodulatory treatment) are beneficial for patients with triple sero-positivity antiphospholipid carriers.

2. **Low dose aspirin with low molecular weight heparin (LMWH):** Patients with obstetric morbidness not fulfilling benchmark for APS have similar pregnancy outcomes to those patients with obstetric manifestations and shows beneficial outcomes after treatment with LDASA in combination with LMWH [63].

3. **Hydroxychloroquine (HCQ):** HCQ possesses the immunomodulatory properties and have been advised to be useful for pregnant women with antiphospholipid [64, 65]. Several lines of evidence based on experimental and animal studies showed that it is capable of antagonizing the antiphospholipid mediated inhibition of formation of trophoblast [64, 66, 67]. A retrospective study based on clinical studies validated the potency of HCQ in obstetric APS which is unmanageable with conventional treatment [68–70]. A randomized, placebo-controlled clinical trial (HYPATIA) proved the effectiveness of HCQ in treatment of pregnant antiphospholipid carriers [71]. Another concern about antiphospholipid sero-positive patients is the puerperium which is a high risk time-period for thrombosis.

5. Future perspectives: research agenda

Recurrent pregnancy loss is a crucial reproductive health problem globally. Several risk factors have been reported in the studies and effective treatment strategies are developed. True prevalence of recurrent miscarriage must be reported globally. Multicentre and international clinical trial studies must be carried out to identify the subgroups that get benefit from treatment. Further, unveiling the mechanism behind endometrial homeostasis and inter-cycle variability during decidual responses regarding recurrent miscarriage as well as determining whether the pregnancy loss itself impacts succeeding endometrial remodeling and placental function may enhance the understanding about endometrial homeostasis. Additionally, the prospective case–control studies are required in which individuals tested +ve for anti-β2GPI antibody to confirm that it can act as a most frequently detected biomarker in association with anti-HLA-DR for recurrent pregnancy loss.

Author details

Nilam Bhasker
Department of Pathology, ESIC, Lucknow, India

*Address all correspondence to: nilam2711@gmail.com

References

[1] ESHRE Guideline Group on RPL, Bender Atik R, Christiansen OB, Elson J, Kolte AM, Lewis S, Middeldorp S, Nelen W, Peramo B, Quenby S, Vermeulen N. ESHRE guideline: recurrent pregnancy loss. Human reproduction open. 2018;2018(2):hoy004.

[2] Practice Committee of the American Society for Reproductive Medicine. Evaluation and treatment of recurrent pregnancy loss: a committee opinion. Fertility and sterility. 2012 Nov 1;98(5): 1103-1111.

[3] Regan L, Braude PR, Trembath PL. Influence of past reproductive performance on risk of spontaneous abortion. British Medical Journal. 1989 Aug 26;299(6698):541-545.

[4] Baek KH, Lee EJ, Kim YS. Recurrent pregnancy loss: the key potential mechanisms. Trends in molecular medicine. 2007 Jul 1;13(7):310-317.

[5] Wilcox AJ, Weinberg CR, O'Connor JF, Baird DD, Schlatterer JP, Canfield RE, Armstrong EG, Nisula BC. Incidence of early loss of pregnancy. New England Journal of Medicine. 1988 Jul 28;319(4):189-194.

[6] Alijotas-Reig J, Garrido-Gimenez C. Current concepts and new trends in the diagnosis and management of recurrent miscarriage. Obstetrical & gynecological survey. 2013 Jun 1;68(6):445-466.

[7] van Dijk MM, Kolte AM, Limpens J, Kirk E, Quenby S, van Wely M, Goddijn M. Recurrent pregnancy loss: diagnostic workup after two or three pregnancy losses? A systematic review of the literature and meta-analysis. Human reproduction update. 2020 Apr 15;26(3): 356-367.

[8] Chard T. 11 Frequency of implantation and early pregnancy loss in natural cycles. Bailliere's clinical obstetrics and gynaecology. 1991 Mar 1;5(1):179-189.

[9] Wilcox AJ, Weinberg CR, O'Connor JF, Baird DD, Schlatterer JP, Canfield RE, Armstrong EG, Nisula BC. Incidence of early loss of pregnancy. New England Journal of Medicine. 1988 Jul 28;319(4):189-194.

[10] Evdokia D, Ellen M, Shigeru S, Kutteh WH, Brosens JJ. Recurrent pregnancy loss (Primer). Nature Reviews: Disease Primers. 2020;6(1).

[11] Ammon Avalos L, Galindo C, Li DK. A systematic review to calculate background miscarriage rates using life table analysis. Birth Defects Research Part A: Clinical and Molecular Teratology. 2012 Jun;94(6):417-423.

[12] Pandey MK, Rani R, Agrawal S. An update in recurrent spontaneous abortion. Archives of gynecology and obstetrics. 2005 Aug;272(2):95-108.

[13] Rai R, Regan L. Recurrent miscarriage. The lancet. 2006 Aug 12;368(9535):601-611.

[14] Christiansen OB, Andersen AM, Bosch E, Daya S, Delves PJ, Hviid TV, Kutteh WH, Laird SM, Li TC, van der Ven K. Evidence-based investigations and treatments of recurrent pregnancy loss. Fertility and sterility. 2005 Apr 1;83(4):821-839.

[15] Pereza N, Ostojić S, Kapović M, Peterlin B. Systematic review and meta-analysis of genetic association studies in idiopathic recurrent spontaneous abortion. Fertility and sterility. 2017 Jan 1;107(1):150-159.

[16] Kwak-Kim J, Skariah A, Wu L, Salazar D, Sung N, Ota K. Humoral and cellular autoimmunity in women with recurrent pregnancy losses and repeated implantation failures: a possible role of vitamin D. Autoimmunity reviews. 2016 Oct 1;15(10):943-947.

[17] Carp HJ, Selmi C, Shoenfeld Y. The autoimmune bases of infertility and pregnancy loss. Journal of autoimmunity. 2012 May 1;38(2-3): J266-J274.

[18] Rai RS, Clifford K, Cohen H, Regan L. High prospective fetal loss rate in untreated pregnancies of women with recurrent miscarriage and antiphospholipid antibodies. Human reproduction. 1995 Dec 1;10(12): 3301-3304.

[19] Chighizola CB, Borghi MO, Meroni PL. Antibodies and diagnostic tests in antiphosholipid syndrome. InSystemic Lupus Erythematosus 2021 Jan 1 (pp. 565-574). Academic Press.

[20] Chighizola CB, Raschi E, Borghi MO, Meroni PL. Update on the pathogenesis and treatment of the antiphospholipid syndrome. Current opinion in rheumatology. 2015 Sep 1;27(5):476-482.

[21] Agostinis C, Biffi S, Garrovo C, Durigutto P, Lorenzon A, Bek A, Bulla R, Grossi C, Borghi MO, Meroni P, Tedesco F. In vivo distribution of β2 glycoprotein I under various pathophysiologic conditions. Blood, The Journal of the American Society of Hematology. 2011 Oct 13;118(15): 4231-4238.

[22] Poulton K, Ripoll VM, Pericleous C, Meroni PL, Gerosa M, Ioannou Y, Rahman A, Giles IP. Purified IgG from Patients with Obstetric but not IgG from Non-obstetric Antiphospholipid Syndrome Inhibit Trophoblast Invasion. American Journal of Reproductive Immunology. 2015 May;73(5):390-401.

[23] Viall CA, Chamley LW. Histopathology in the placentae of women with antiphospholipid antibodies: a systematic review of the literature. Autoimmunity reviews. 2015 May 1;14(5):446-471.

[24] Abrahams VM, Chamley LW, Salmon JE. Antiphospholipid syndrome and pregnancy: pathogenesis to translation. Arthritis & rheumatology (Hoboken, NJ). 2017 Sep;69(9):1710.

[25] Rand JH, Wu XX, Guller S, Gil J, Guha A, Scher J, Lockwood CJ. Reduction of annexin-V (placental anticoagulant protein-I) on placental villi of women with antiphos pholipid antibodies and recurrent spontaneous abortion. American journal of obstetrics and gynecology. 1994 Dec 1;171(6): 1566-1572.

[26] Salafia CM, Cowchock FS. Placental pathology and antiphospholipid antibodies: a descriptive study. American journal of perinatology. 1997 Sep;14(08): 435-441.

[27] Sebire NJ, Fox H, Backos M, Rai R, Paterson C, Regan L. Defective endovascular trophoblast invasion in primary antiphospholipid antibody syndrome-associated early pregnancy failure. Human Reproduction. 2002 Apr 1;17(4):1067-1071.

[28] Bhasker N, Kar M, Kumar D. Estimation of Antiphospholipid Antibodies, Anticardiolipin Antibodies, Beta-2 Glycoprotein 1 Antibodies in Recurrent Pregnancy Loss-A Case-control Study. Journal of Clinical & Diagnostic Research. 2018 Oct 1;12(10).

[29] Rai RS, Regan L, Clifford K, Pickering W, Dave M, Mackie I,

McNally T, Cohen H. Immunology: Antiphospholipid antibodies and β2-glycoprotein-I in 500 women with recurrent miscarriage: results of a comprehensive screening approach. Human Reproduction. 1995 Aug 1;10(8):2001-2005.

[30] Yetman DL, Kutteh WH. Antiphospholipid antibody panels and recurrent pregnancy loss: prevalence of anticardiolipin antibodies compared with other antiphospholipid antibodies. Fertility and sterility. 1996 Oct 1;66(4):540-546.

[31] Alijotas-Reig J, Esteve-Valverde E, Ferrer-Oliveras R, Sáez-Comet L, Lefkou E, Mekinian A, Belizna C, Ruffatti A, Tincani A, Marozio L, Espinosa G. The European Registry on Obstetric Antiphospholipid Syndrome (EUROAPS): a survey of 1000 consecutive cases. Autoimmunity reviews. 2019 Apr 1;18(4):406-414.

[32] Bates SM, Greer IA, Middeldorp S, Veenstra DL, Prabulos AM, Vandvik PO. VTE, thrombophilia, antithrombotic therapy, and pregnancy: antithrombotic therapy and prevention of thrombosis: American College of Chest Physicians Evidence-Based Clinical Practice Guidelines. Chest. 2012 Feb 1;141(2): e691S-e736S.

[33] Bouvier S, Cochery-Nouvellon E, Lavigne-Lissalde G, Mercier E, Marchetti T, Balducchi JP, Mares P, Gris JC. Comparative incidence of pregnancy outcomes in treated obstetric antiphospholipid syndrome: the NOH-APS observational study. Blood, The Journal of the American Society of Hematology. 2014 Jan 16;123(3):404-413.

[34] Chamley LW, Duncalf AM. Action of anticardiolipin and antibodies to beta 2-glycoprotein-I on trophoblast proliferation as a mechanism for fetal death. The Lancet. 1998;352(9133): 1037-1038.

[35] Regal JF, Gilbert JS, Burwick RM. The complement system and adverse pregnancy outcomes. Molecular immunology. 2015 Sep 1;67(1):56-70.

[36] Salmon JE, Girardi G, Holers VM. Activation of complement mediates antiphospholipid antibody-induced pregnancy loss. Lupus. 2003 Jul;12(7):535-538.

[37] Pierangeli SS, Girardi G, Vega-Ostertag M, Liu X, Espinola RG, Salmon J. Requirement of activation of complement C3 and C5 for antiphospholipid antibody–mediated thrombophilia. Arthritis & Rheumatism. 2005 Jul;52(7):2120-2124.

[38] Green DM, O'Donoghue K. A review of reproductive outcomes of women with two consecutive miscarriages and no living child. Journal of Obstetrics and Gynaecology. 2019 Aug 18;39(6):816-821.

[39] Jeve Y, Rana R, Bhide A, Thangaratinam S. Accuracy of first-trimester ultrasound in the diagnosis of early embryonic demise: a systematic review. Ultrasound in obstetrics & gynecology. 2011 Nov;38(5):489-496.

[40] Brown DL, Packard A, Maturen KE, Deshmukh SP, Dudiak KM, Henrichsen TL, Meyer BJ, Poder L, Sadowski EA, Shipp TD, Simpson L. ACR Appropriateness Criteria® first trimester vaginal bleeding. Journal of the American College of Radiology. 2018 May 1;15(5):S69-S77.

[41] National Collaborating Centre for Women's and Children's Health (UK. Ectopic pregnancy and miscarriage: diagnosis and initial management in early pregnancy of ectopic pregnancy and miscarriage.

[42] Preisler J, Kopeika J, Ismail L, Vathanan V, Farren J, Abdallah Y, Battacharjee P, Van Holsbeke C, Bottomley C, Gould D, Johnson S. Defining safe criteria to diagnose miscarriage: prospective observational multicentre study. Bmj. 2015 Sep 23;351.

[43] Pillai RN, Konje JC, Tincello DG, Potdar N. Role of serum biomarkers in the prediction of outcome in women with threatened miscarriage: a systematic review and diagnostic accuracy meta-analysis. Human reproduction update. 2016 Mar 1;22(2):228-239.

[44] Ghazeeri GS, Kutteh WH. Immunological testing and treatment in reproduction: frequency assessment of practice patterns at assisted reproduction clinics in the USA and Australia. Human Reproduction. 2001 Oct 1;16(10): 2130-2135.

[45] Segawa T, Kuroda T, Kato K, Kuroda M, Omi K, Miyauchi O, Watanabe Y, Okubo T, Osada H, Teramoto S. Cytogenetic analysis of the retained products of conception after missed abortion following blastocyst transfer: a retrospective, large-scale, single-centre study. Reproductive biomedicine online. 2017 Feb 1;34(2):203-210.

[46] Sugiura-Ogasawara M, Aoki K, Fujii T, Fujita T, Kawaguchi R, Maruyama T, Ozawa N, Sugi T, Takeshita T, Saito S. Subsequent pregnancy outcomes in recurrent miscarriage patients with a paternal or maternal carrier of a structural chromosome rearrangement. Journal of human genetics. 2008 Jul 1;53(7):622-628.

[47] Morita K, Ono Y, Takeshita T, Sugi T, Fujii T, Yamada H, Nakatsuka M, Fukui A, Saito S. Risk factors and outcomes of recurrent pregnancy loss in Japan. Journal of Obstetrics and Gynaecology Research. 2019 Oct;45(10):1997-2006.

[48] Franssen MT, Korevaar JC, van der Veen F, Leschot NJ, Bossuyt PM, Goddijn M. Reproductive outcome after chromosome analysis in couples with two or more miscarriages: case-control study. Bmj. 2006 Mar 30;332(7544):759-763.

[49] Zhang Y, Lei Q, Liu J, Lin M, Luo L, Li T, Wang Q, Zhou C. Selective growth of mosaic cells in chromosomal analysis of chorionic villi by conventional karyotyping. Molecular and cellular probes. 2020 Jun 1;51:101532.

[50] Maisenbacher MK, Merrion K, Kutteh WH. Single-nucleotide polymorphism microarray detects molar pregnancies in 3% of miscarriages. Fertility and sterility. 2019 Oct 1;112(4):700-706.

[51] Hahnemann JM, Vejerslev LO. European collaborative research on mosaicism in CVS (EUCROMIC)—fetal and extrafetal cell lineages in 192 gestations with CVS mosaicism involving single autosomal trisomy. American journal of medical genetics. 1997 May 16;70(2):179-187.

[52] Kudesia R, Li M, Smith J, Patel A, Williams Z. Rescue karyotyping: a case series of array-based comparative genomic hybridization evaluation of archival conceptual tissue. Reproductive Biology and Endocrinology. 2014 Dec;12(1):1-8.

[53] Redin C, Brand H, Collins RL, Kammin T, Mitchell E, Hodge JC, Hanscom C, Pillalamarri V, Seabra CM, Abbott MA, Abdul-Rahman OA. The genomic landscape of balanced cytogenetic abnormalities associated with human congenital anomalies. Nature genetics. 2017 Jan;49(1):36-45.

[54] Sahoo T, Dzidic N, Strecker MN, Commander S, Travis MK, Doherty C, Tyson RW, Mendoza AE, Stephenson M, Dise CA, Benito CW. Comprehensive genetic analysis of pregnancy loss by chromosomal microarrays: outcomes, benefits, and challenges. Genetics in medicine. 2017 Jan;19(1):83-89.

[55] Popescu F, Jaslow CR, Kutteh WH. Recurrent pregnancy loss evaluation combined with 24-chromosome microarray of miscarriage tissue provides a probable or definite cause of pregnancy loss in over 90% of patients. Human Reproduction. 2018 Apr 1;33(4):579-587.

[56] Papas RS, Kutteh WH. A new algorithm for the evaluation of recurrent pregnancy loss redefining unexplained miscarriage: review of current guidelines. Current Opinion in Obstetrics and Gynecology. 2020 Oct 1;32(5):371-379.

[57] Miyakis S, Lockshin MD, Atsumi T, Branch DW, Brey RL, Cervera RH, Derksen RH, De Groot PG, Koike T, Meroni PL, Reber G. International consensus statement on an update of the classification criteria for definite antiphospholipid syndrome (APS). Journal of thrombosis and haemostasis. 2006 Feb;4(2):295-306.

[58] Lakos G, Favaloro EJ, Harris EN, Meroni PL, Tincani A, Wong RC, Pierangeli SS. International Consensus Guidelines on Anticardiolipin and Anti–2-Glycoprotein I Testing. Arthritis & Rheumatism. 2012 Jan;64(1):1-0.

[59] Passam FH, Krilis SA. Laboratory tests for the antiphospholipid syndrome: current concepts. Pathology. 2004 Apr 1;36(2):129-138.

[60] Hunt JE, McNeil HP, Morgan GJ, Crameri RM, Krilis SA. A phospholipid-β2-glycoprotein I complex is an antigen for anticardiolipin antibodies occurring in autoimmune disease but not with infection. Lupus. 1992 Feb;1(2):75-81.

[61] Harris EN, Gharavi AE, Hughes GR. Anti-phospholipid antibodies. Clinics in rheumatic diseases. 1985 Dec 1;11(3): 591-609.

[62] Smith LJ. Laboratory Diagnosis of the Lupus Anticoagulant. American Society for Clinical Laboratory Science. 2017 Jan 1;30(1):7-14.

[63] Giannakopoulos B, Passam F, Ioannou Y, Krilis SA. How we diagnose the antiphospholipid syndrome. Blood, The Journal of the American Society of Hematology. 2009 Jan 29;113(5): 985-994.

[64] Moore GW. Recent guidelines and recommendations for laboratory detection of lupus anticoagulants. InSeminars in thrombosis and hemostasis 2014 Mar (Vol. 40, No. 02, pp. 163-171). Thieme Medical Publishers.

[65] Roberge S, Giguère Y, Villa P, Nicolaides K, Vainio M, Forest JC, von Dadelzen P, Vaiman D, Tapp S, Bujold E. Early administration of low-dose aspirin for the prevention of severe and mild preeclampsia: a systematic review and meta-analysis. Obstetrical & gynecological survey. 2012 Dec 1;67(12):760-762.

[66] Erkan D, Patel S, Nuzzo M, Gerosa M, Meroni PL, Tincani A, Lockshin MD. Management of the controversial aspects of the antiphospholipid syndrome pregnancies: a guide for clinicians and researchers. Rheumatology. 2008 Jun 1;47(suppl_3): iii23-7.

[67] Del Ross T, Ruffatti A, Visentin MS, Tonello M, Calligaro A, Favaro M, Hoxha A, Punzi L. Treatment of 139 pregnancies in antiphospholipid-positive

women not fulfilling criteria for antiphospholipid syndrome: a retrospective study. The Journal of rheumatology. 2013 Apr 1;40(4): 425-429.

[68] Soh MC, Pasupathy D, Gray G, Nelson-Piercy C. Persistent antiphospholipid antibodies do not contribute to adverse pregnancy outcomes. Rheumatology. 2013 Sep 1;52(9):1642-1647.

[69] Agmon-Levin N, Blank M, Zandman-Goddard G, Orbach H, Meroni PL, Tincani A, Doria A, Cervera R, Miesbach W, Stojanovich L, Barak V. Vitamin D: an instrumental factor in the anti-phospholipid syndrome by inhibition of tissue factor expression. Annals of the Rheumatic Diseases. 2011 Jan 1;70(1):145-150.

[70] Alijotas-Reig J, Ferrer-Oliveras R. The European Registry on Obstetric Antiphospholipid Syndrome (EUROAPS): a preliminary first year report. Lupus. 2012 Jun;21(7):766-768.

[71] Meroni PL. Prevention & treatment of obstetrical complications in APS: Is hydroxychloroquine the Holy Grail we are looking for?. Journal of autoimmunity. 2016 Dec 1;75:1-5.